Multiplica
the Algebra

Authors

Dr. Arthur Wiebe
Sheldon Erickson
Cheryl Hartshorn
Michelle Pauls

Illustrator

Brenda Wood

Technical Illustrator

Michelle Pauls

Editors

Betty Cordel
Michelle Pauls

Desktop Publishing

Tanya Adams
Tracey Lieder
Leticia Rivera

Developed and Published
by

AIMS Education Foundation

This book contains materials developed by the AIMS Education Foundation. **AIMS** (**A**ctivities **I**ntegrating **M**athematics and **S**cience) began in 1981 with a grant from the National Science Foundation. The non-profit AIMS Education Foundation publishes hands-on instructional materials that build conceptual understanding. The foundation also sponsors a national program of professional development through which educators may gain expertise in teaching math and science.

AIMS Education Foundation
1595 S. Chestnut Ave., Fresno, CA 93702
888.733.2467 • aimsedu.org

ISBN **978-1-60519-084-6**

Printed in the United States of America

Table of Contents

I Hear and I Forget,
I See and I Remember,
I Do and I Understand.

– Chinese Proverb

The Distributive Property

The distributive property of multiplication over addition belongs to the ABCs of mathematics. It plays a central role in arithmetic and algebra and, therefore, deserves special attention.

A deep understanding of the distributive property *requires that students thoroughly understand important number base concepts.* This is especially true for middle school students preparing for the formal study of algebra. It may come as new insight that key number base concepts are imbedded in the "unknown *x*" of algebra. All of this will become evident to students as they work through the activities in this publication.

The distributive property algorithms used in algebra differ sharply from the standard multiplication algorithm used in arithmetic. The standard arithmetic algorithm, while efficient for generating answers, is seriously deficient in several important respects: it masks the ten-ness of our numeration system, does not require thinking in terms of base ten, consists of a series of mysterious steps, and *does not transfer into algebra.*

In these activities, students are introduced to algebraic concepts and methods of multiplication while still in the familiar environment of the base-ten system. While in that environment, students must constantly attend to the "ten-ness" of our numeration system. They will find that the new algorithms are rich in meaning and devoid of mystery. Once familiar with application of these concepts and processes in base ten, students transfer their understandings to other number bases where they learn to think in terms of "three-ness," "four-ness," "five-ness," etc. As a result, they become aware of the universal nature of the processes. The journey culminates as students transfer all of their understandings and experience to thinking in general terms, or what might be dubbed "*x*-ness."

If students are to be prepared for algebra, they must master the algorithms used in these activities. The first algorithm that is introduced is known variously as *display multiplication* or *partial product multiplication*. In it, the process of multiplication is broken down into its components. This clarifies what is going on within the process. Display multiplication transfers intact into algebra and later mathematics and, therefore, should be introduced as early as possible. In fact, many newer texts strongly advocate that students learn this algorithm before they are introduced to the standard algorithm.

In the early activities, the distributive property is studied in parallel numeric and geometric contexts, making them suitable for students beginning in grade five. At this stage, students develop understanding and facility with these forms. Later, students transfer these intact to algebraic expressions and equations using parallel numeric, geometric, and algebraic forms. Physical models and geometric representations facilitate student understanding as they encounter algebraic operations and expressions. Through this process, students bring an understanding of and facility with key mathematical concepts associated with multiplication to their study of algebra.

The activities throughout the publication utilize manipulatives and representations to help students *build mental images* of imbedded concepts and processes. Firm connections are developed among physical layouts, representational pictures, numerical expressions, and algebraic expressions. AIMS Base Ten Blocks are the manipulatives of choice for studies in base ten and AIMS Algebra Tiles are the choice for studies in bases three, four, five, and *x*.

AIMS Algebra and the Cartesian Coordinate Plane

A hallmark of the AIMS Algebra Program is the consistency with which it models multiplication using manipulatives placed on the Cartesian coordinate plane. Since such modeling is an integral component of most activities, it is important for users to recognize the rationale behind its selection and the rich contribution it makes to student understanding. Before using these activities with students, it is essential to gain a thorough understanding of the model and its usage.

At the outset, it is important for users to recognize that the AIMS approach avoids a common pitfall inherent in those programs whose placement of tiles could easily be construed as taking place in the fourth quadrant. While such programs do not associate their placement of tiles with the coordinate plane, their usage could easily lead to confusion as students encounter standard algebraic representation which is on the coordinate plane. By using the coordinate plane as the "playing field" for modeling, AIMS Algebra distinguishes itself from such programs. *AIMS avoids confusion by being consistent with algebraic usage in the placement of tiles on the coordinate plan as models are constructed.*

A New Model for Algebraic Expressions and Equations

Students are more likely to gain an in-depth understanding of algebraic concepts and processes when they are studied at concrete, representational, and abstract levels simultaneously. Such a comprehensive approach utilizes physical models, sketches of models, and written algebraic statements interpreting the models. Each form makes its own significant contribution. Understanding deepens as students compare and contrast the forms.

This approach creates the need for a model that accurately reflects algebraic concepts and processes. Fortunately, such a model has recently been introduced by mathematics educator and author Behrouz B. Aghevli. His model exhibits these essential features:

1. models of algebraic expressions are constructed on the Cartesian coordinate plane, which imparts meaning to the components;
2. pictures of algebraic expressions are sketched on the coordinate plane;
3. positive and negative terms derive meaning from their placement on the coordinate plane;
4. algebraic processes are fully and accurately modeled;
5. algebraic processes take on a one-to-one correspondence among all three representations; and
6. first, second, and third degree expressions and equations can be modeled and pictured. Pictures of third degree expressions involve isometric drawings.

Because of these superior features, AIMS Algebra has adopted this model for its program.

Standard interpretations of the coordinate plane form the basis for imparting meaning. Consider four points identified by ordered pairs, one in each quadrant.

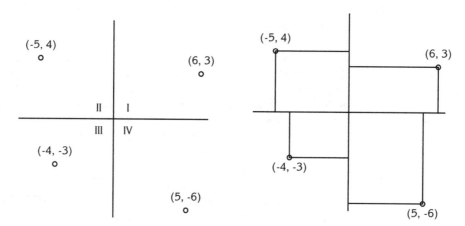

Each point can serve as the vertex of a rectangle whose opposite vertex is at the origin. In such rectangles, the ordered pairs tell us their length and width. Areas are found by multiplying length times width. The area in the first quadrant is 6 x 3 = 18 square units; in the second quadrant it is -5 x 4 = -20 square units; in the third quadrant it is -4 x -3 =12 square units; and in the fourth quadrant it is 5 x -6 = -30 square units.

Note that rectangles in quadrants I and III have positive areas and those in quadrants II and IV have negative areas.

The idea of a negative area may be new to students. They need to know that components of a model are frequently *assigned meanings arbitrarily*. In this instance, we conclude that an area is negative or positive based on the argument that placement on the coordinate plane justifies this interpretation. The justification is based on the fact that the product of any ordered pair for any point inside or on the border of any rectangle will always be positive in quadrants I and III and negative in quadrants II and VI.

Positive and negative areas can represent a broad range of ideas. For example, areas in quadrants I and III can be thought of as regions with surplus water, whereas areas in quadrants II and IV have equal water deficiencies. Permit the water to flow between them and equal areas with equal surpluses and deficiencies will negate their deviations from normal just as equal positive and negative numbers added together negate each other to equal zero.

The model used in this discussion utilizes four components: small cubes representing units or 1, sticks representing lengths of n units, flats representing areas of n^2 square units, and large cubes representing volumes of n^3 cubic units.

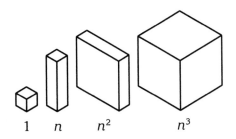

$$1 \qquad n \qquad n^2 \qquad n^3$$

The following are examples of the placement of these components on the coordinate plane to represent various algebraic expressions.

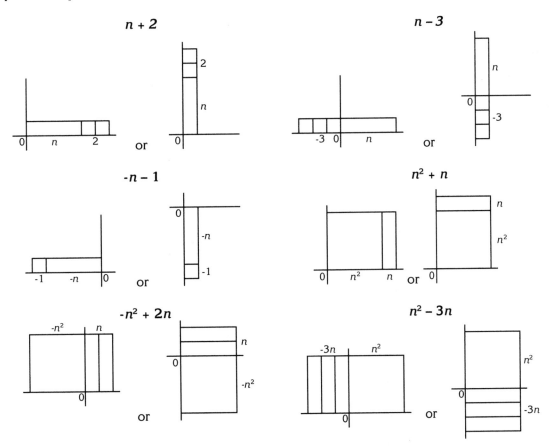

In summary, both dimensions in the first quadrant are positive so their product is positive and regions are considered to be positive. In the second and fourth quadrant, one dimension is positive and the other negative so their product is negative and regions are considered to be negative. In the third quadrant, both dimensions are negative so their product is positive and regions are considered to be positive.

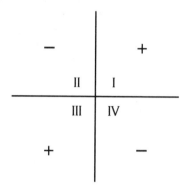

Multiplication in graphical form consists of constructing a rectangle with one factor graphed as the first side and the second factor as an adjoining side drawn perpendicular to the first. In the graphical display of $(n-2)(n+1)$, $n-2$ is here graphed on the horizontal axis and $(n+1)$ on the vertical axis. Note that *the outer extremities of each factor's graph on the axes serve as boundary markers for the rectangle* shown by the dashed line.

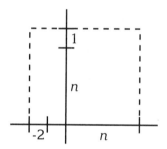

Part of the above rectangle lies left of the vertical axis and part to the right. The *left part is a negative region.* Every point in that region is the product of a positive and a negative number. *The part to the right of the vertical axis is a positive region.* Every point in that region is the product of two positive numbers.

Once the rectangle is outlined, the vertical and horizontal *breaks* between components are drawn *through the entire rectangle.* Every break between components on either axis becomes a point from which a line is drawn through the rectangle.

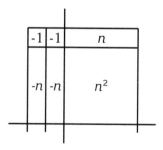

This rectangle is now made up of six components in four groups. Four components lie in negative territory and two in positive territory. The large n^2 flat is positive. Of the three sticks or n-tiles, one is positive and two are negative. The two unit cubes are negative. These four groups mirror the algebraic product obtained using the distributive property of multiplication.

The results are simplified by collecting terms. Just as at the abstract level, this is achieved with manipulatives by pairing equal positive and negative components and eliminating such pairs. Only one such pair exists in this example: one positive and one negative n-tile. When this pair is removed, a single negative n-tile remains.

Initial equation: $(n-2)(n+1) = n^2 + n - 2n - 2$
Equation after collection of terms: $(n-2)(n+1) = n^2 - n - 2$

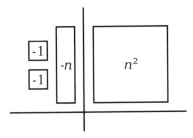

The general sequence, with attendant rules, is as follows. A rectangle with dimensions of $(n+3)$ and $(n-2)$ is used as an example in the following. The components before simplification are: 1 positive n^2-flat, 3 positive and 2 negative n-sticks, and 6 negative units.

1. Select the indicated number of each component type. Place negative components in the second quadrant and positive components in the first quadrant of the coordinate plane.
2. Write an initial algebraic statement identifying the number, type, and sign of each component subset.
3. Form the components into a rectangle, keeping each in the proper quadrant.

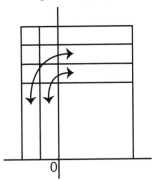

4. Sketch a picture of the completed rectangle.
5. From the model and/or picture, determine the dimensions of the rectangle.
6. Write an equation in the form *(factor)(factor) = product* with terms uncollected as in this example: $(n+3)(n-2) = n^2 - 2n + 3n - 6$.
7. In the model, simplify the product by removing all congruent positive and negative pairs. In parallel fashion collect the terms in the product of the equation and write it in simplified form:

Initial equation: $(n+3)(n-2)$ $= n^2 - 2n + 3n - 6.$
Equation after simplification: $= n^2 + n - 6$

The initial activities that follow provide practice in the simplification of terms at the manipulative, representational, and abstract levels.

Building a Model for Multiplication

The activities in *Part One* provide students with a hands-on introduction to the model for multiplication used in this publication. From the outset, students use base ten tiles to explore the relationships among unit squares, sticks, and flats and how these are used in building models of multiplication. Numerous key concepts and skills foundational to algebra begin to emerge during these explorations. The hands-on approach enriches the learning environment. Seeing concepts and skills in action at the manipulative level helps to build understanding.

Sufficient time, multiple experiences, and frequent strategic questioning of student understanding are essential if students are to gain the maximum benefit from these early experiences. If the foundation is well developed, students will progress through the subsequent activities more rapidly and with greater benefit.

A brief overview of each activity follows.

Multiplying with Tens

Students familiarize themselves with the base ten manipulatives and their interrelationships; calculate the number of unit squares in a set of base ten tiles by using multiples of ten; and identify each base ten tile with its value as a multiple of ten. Place value is reinforced throughout. Mastery of place value and its role in multiplication is essential if students are to meaningfully use a multiplication strategy.

Students follow a three-step process: constructing a model of the multiplication problem, drawing a sketch of the model, and determining the numerical product.

Building Rectangles

Students learn to model multiplication by constructing arrays; use powers of ten to count large arrays; and recognize patterns within arrays. Students construct arrays by filling rectangles with the fewest number of pieces in a prescribed order.

In the area model, the two factors give the dimensions of the rectangle, and the number of unit squares it takes to fill the rectangle gives the product.

Picturing Rectangles

Students determine the solution of a multiplication problem by finding the sum of the partial products and learn to solve multi-digit multiplication problems with arrays. They come to realize that the product of a pair of two-digit non-zero numbers always consists of four regions: the largest representing hundreds; two smaller regions representing tens; and the smallest consisting of units or ones.

By asking students to construct the rectangle from the dimensions, this activity forces students to convert numerical information into representational form. Practice at the representational level reinforces processes used at the abstract level.

Writing Rectangles

Students learn to solve multi-digit multiplication problems with a partial product algorithm often referred to as the display multiplication algorithm. The partial product algorithm provides a meaningful solution connected to concrete and representational models while retaining the place value meaning of the digits. It encourages students to practice multiplication by tens and develop their estimation capabilities. In the process, they build a mental image of the solution.

Multiplying with TENS

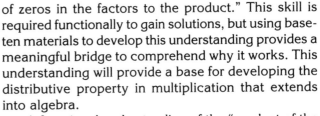

Topic
Estimation by using multiples of ten

Key Question
How can you determine the amount of unit cubes in a set of base-ten blocks?

Learning Goals
Students will:
- be able to calculate the number of unit cubes in a set of base-ten blocks by using multiples of ten, and
- identify each base-ten block with its value as a multiple of ten.

Guiding Documents
Project 2061 Benchmarks
- *Add, subtract, multiply, and divide whole numbers mentally, on paper, and with a calculator.*
- *Express numbers like 100, 1000, and 1,000,000 as powers of 10.*

Common Core State Standards for Mathematics*
- *Construct viable arguments and critique the reasoning of others. (MP.3)*
- *Look for and make use of structure. (MP.7)*
- *Understand the place value system. (5.NBT.A)*

Math
Multiplication
Place value
Estimation

Integrated Processes
Observing
Comparing and contrasting
Generalizing

Materials
Base-ten blocks

Background Information
Mastery of place value and its effects on multiplying is required for students to meaningfully use a multiplication strategy. It is the basic component of all multiplication algorithms and provides the ability to estimate solutions to multiplication problems.

Students often memorize an algorithm for multiplying multiples of ten such as: "use the product of significant figures in the factors and add the number of zeros in the factors to the product." This skill is required functionally to gain solutions, but using base-ten materials to develop this understanding provides a meaningful bridge to comprehend why it works. This understanding will provide a base for developing the distributive property in multiplication that extends into algebra.

A functional understanding of the "product of the significant figures plus the zeros of the factors" can be enhanced by looking at the process in scientific notation.

$$2000 \quad \times \quad 30{,}000 \quad = \quad 60{,}000{,}000$$
$$\text{(6 with 7 zeros)}$$
$$(2 \times 10^3) \quad \times \quad (3 \times 10^4) \quad = \quad (6 \times 10^7)$$

When the process is viewed in both standard and scientific forms, the adding of the zeros is really the collection of the exponents developed in the rules of exponents.

Management
1. The pace of this lesson varies greatly with the experience and age of the students. Older, more-experienced students will move rapidly away from the manipulative and representational levels to the numeric form. Younger, less-experienced students will make the transition more slowly and may need to depend on the materials throughout the lesson.
2. This is an introductory lesson for students who have little experience with multiplying numbers by powers of ten.

Procedure
1. To have students become familiar with the base ten materials, have them get out at least 10 unit cubes, 10 sticks, and a flat.
2. Direct the students to trace the outline of a unit cube and a stick on paper and write number sentences of the observations that are made as the teacher guides the following discussion:
   a. Each unit cube will cover one square. How many squares will a stick cover? [one stick = 10 squares]
   b. How many cubes wide and long is a stick? [one cube wide, 10 cubes long]
   c. How can you use the length and width dimensions to determine the number of cubes in a stick? [one cube x 10 cubes = 10 cubes]

3. Have the students trace the outline of a flat onto their papers and write number sentences of the observations that are made as the teacher guides the following discussion:
   a. How many stick lengths wide is a flat? [one stick]
   b. How many sticks laid next to each other does it take to cover a flat? [10 sticks = one flat]
   c. How many unit cubes long is a flat? [10 cubes]
   d. How many unit cubes wide is a flat? [10 cubes]
   e. How many unit cubes will a flat cover? [one flat = 100 squares, 10 cubes x 10 cubes = 100 cubes]
4. After distributing the record sheet, direct the students to build rectangles based on one set of facts.
5. Have the students draw sketches and record the dimensions of each of their solutions.
6. After a variety of fact sets have been explored, lead a discussion of the patterns of multiples of ten they observed. [Significant figures in the solution are the product of significant figures in the factors. The number of zeros in the factors are equal to the number of zeros in the product.]
7. Give the students situations to solve using the base-ten materials. Direct them to construct the solutions, to sketch them, and to write number sentences relating the dimensions of the rectangles to the areas. Some examples of situations are: How many unit cubes would be covered by 12 sticks? [1 x 120 = 120, 2 x 60 = 120, (3 x 40 = 120,) 4 x 30 = 120, 6 x 20 = 120, 12 x 10 = 120] How many cubes would be covered by a rectangle made of six flats? [10 x 60 = 600, 20 x 30 = 600]
8. Have the students suggest multiplication problems of two factors that are multiples of ten and determine their solution using manipulatives, sketches, or abstractly, but direct them to record their solutions in number sentences.

Connecting Learning
1. What patterns do you see in these problems?
2. How could you use the patterns you recognized today to estimate the answers to large multiplication problems? [round factors to largest significant figure's value, multiply rounded figure and add number of zeros in factors]

Extensions
1. Have students estimate a page of multiplication problems using their estimation method and check how close they are with a calculator.
2. If students are familiar with scientific notation, have them convert standard notation problems into scientific notation and see how the process of adding zeros is analogous to adding exponents in scientific notation.

2000	x	30,000	=	60,000,000

(6 with 7 zeros)

$$(2 \times 10^3) \quad x \quad (3 \times 10^4) \quad = \quad (6 \times 10^7)$$

Solutions

ones by ones		ones by tens		tens by tens	
2 x 3	6	2 x 30	60	20 x 30	600
2 x 4	8	2 x 40	80	20 x 40	800
2 x 5	10	2 x 50	100	20 x 50	1000
2 x 6	12	2 x 60	120	20 x 60	1200
2 x 7	14	2 x 70	140	20 x 70	1400
3 x 3	9	3 x 30	90	30 x 30	900
3 x 4	12	3 x 40	120	30 x 40	1200
3 x 5	15	3 x 50	150	30 x 50	1500
4 x 4	16	4 x 40	160	40 x 40	1600

Multiplying with TENS

ones by ones	ones by tens	tens by tens
2 x 3	2 x 30	20 x 30
2 x 4	2 x 40	20 x 40
2 x 5	2 x 50	20 x 50
2 x 6	2 x 60	20 x 60
2 x 7	2 x 70	20 x 70
3 x 3	3 x 30	30 x 30
3 x 4	3 x 40	30 x 40
3 x 5	3 x 50	30 x 50
4 x 4	4 x 40	40 x 40

Multiplying with TENS

Multiplying
Multiplying
Multiplying

Connecting Learning

1. What patterns do you see in these problems?

2. How could you use the patterns you recognized today to estimate the answers to large multiplication problems?

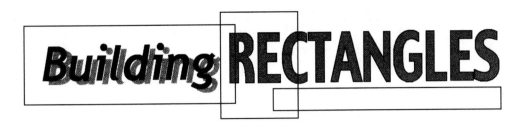

Building RECTANGLES

Topic
Multiplication—area model

Key Question
How can you determine how many base-ten cubes it would take to cover your rectangle?

Learning Goals
Students will:
- learn to model multiplication by constructing an array,
- learn to use powers of ten to count large arrays, and
- recognize patterns of powers of tens within arrays.

Guiding Documents
Project 2061 Benchmarks
- *Add, subtract, multiply, and divide whole numbers mentally, on paper, and with a calculator.*
- *Express numbers like 100, 1000, and 1,000,000 as powers of 10.*

Common Core State Standards for Mathematics*
- *Model with mathematics. (MP.4)*
- *Use appropriate tools strategically. (MP.5)*
- *Look for and make use of structure. (MP.7)*
- *Use place value understanding and properties of operations to perform multi-digit arithmetic. (4.NBT.B)*

Math
Multiplication
Place value
Estimation

Integrated Processes
Observing
Comparing and contrasting
Generalizing

Materials
Base-ten blocks
Butcher paper, cm grid paper, or sample rectangles

Background Information
This investigation has students model the multiplication process in an area array. Prior experience will allow some students to move more quickly than others. Students just being introduced to multiplication will continue with the concrete models for an extended time. Older students will often move directly to the visual record and quickly to a numeric algorithm once the model helps them make sense of prior skills.

In the area model, the two factors give the dimensions of the rectangle, and the number of squares within the rectangle—the area—gives the product. If a rectangle is made from base-ten materials, the squares can be counted rapidly since a flat covers 100 squares, a stick 10 squares, and a cube one square. It is evident that to count the number of squares quickly, the rectangle should be covered with the fewest pieces. This makes it necessary to start with the biggest piece possible. When one factor uses the tens place, sticks will be involved. When both factors have tens in them, flats can be utilized. By being consistent with where rectangles are begun, the patterns become easier to recognize and communicate.

Students may choose a number of ways to record their experience and may switch from one to another as they become more familiar with the process. Young students initially like to make a record on a rectangle that is identical to the sample they receive. They quickly move on to either recording it on a reduced grid; making a box, line, and dot sketch; or developing a rectangle marked into regions.

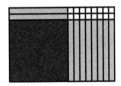

By building and representationally recording a number of rectangles, students will discover some very useful patterns. Visually they will see that flats form a rectangle in the lower left corner. The sticks form rectangles in two regions—a rectangle of sticks placed horizontally in the upper left corner, and a rectangle of sticks placed vertically in the lower right corner. The cubes form a rectangle in the upper right corner. As numbers are added to these representations, numeric patterns can be identified. The flats in the lower left corner are tens by tens products. The two regions of sticks are ones by tens products. The cubes in the upper right corner are the products of the ones place.

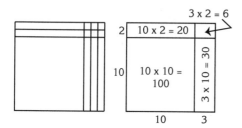

This visual approach provides a very powerful way for many students to make sense of multiplication processes. Some visual learners begin building the rectangles in their minds and only record the products of each region (the partial products) to sum. Other students become so confident of the patterns that they can make box, line, and dot sketches or a rectangle with regions much faster than finding the solution with traditional algorithms.

Management

1. Initially students will want to trace their solutions onto their sample rectangles. Depending on the level of the students, they may choose to make simple box, stick, and dot sketches for their records.
2. The sample rectangles provided are limited by the print space on a page. Larger rectangles can be made by tracing rectangles onto butcher paper, centimeter grid paper that is available in a roll, or by piecing standard sheets together. These can be laminated for repeated use.

Procedure

1. Give students outlines of rectangles (those provided or teacher generated) and discuss strategies for quickly determining how many unit cubes would be needed to cover the rectangle.
2. Provide the students with base-ten materials and direct them to cover their rectangles. Inform them that in order to make comparisons between groups, they should cover the rectangles with the materials using the following guidelines:
   * Fill from the lower left corner where the x is in the circle.
   * Use the fewest number of pieces by filling with flats first, sticks second, and cubes last.
3. Have each student make an individual record of the group's solution on the sample rectangle sheets or with a sketch.

4. Working with their group, direct the students to determine a way to count the squares quickly using their record or the model. Have them record the number of cubes that would be needed to cover their rectangle.
5. When the students have completed a number of rectangles, discuss as a class:
   * their counting methods and
   * patterns of arrangement of flats, sticks, and cubes.
6. Direct the students to return to their record and add the following numeric information:
   * Record the dimensions of the rectangle with tens and ones by the respective pieces of the bottom and left side.
   * Within each region, record the dimensions and product of that region.
   * Within a blank margin, record and find the sum of all the regions' products.
   * Within a blank margin, record the number of cubes and the rectangle's dimensions as a multiplication problem.
7. While the students refer to their records, discuss what visual and numeric patterns they observe emerging.

Connecting Learning

1. What strategies did you use to count up the number of cubes rapidly? [flats by 100, sticks by 10, cubes by one]
2. What patterns did you notice in how the types of pieces were placed in each rectangle? [Flats in a rectangle in the lower left corner, a rectangle of sticks placed horizontally in the upper left corner, a rectangle of sticks placed vertically in the lower right corner, cubes in a rectangle in the upper right corner.]
3. What number patterns do you notice in the record of your solution? [tens by tens problem in lower left corner, ones by tens problem in upper left, tens by ones in lower right corner, and ones by ones problem in upper right corner]
4. How can you use your patterns to help you count the number of cubes in a rectangle if you are only told how long each side is?

Extension

Have students outline rectangles or generate the dimensions of a rectangle on their own. Have them apply their patterns and determine the amount of cubes required to cover the rectangle, and then have them check it by building and recording the model.

Solutions

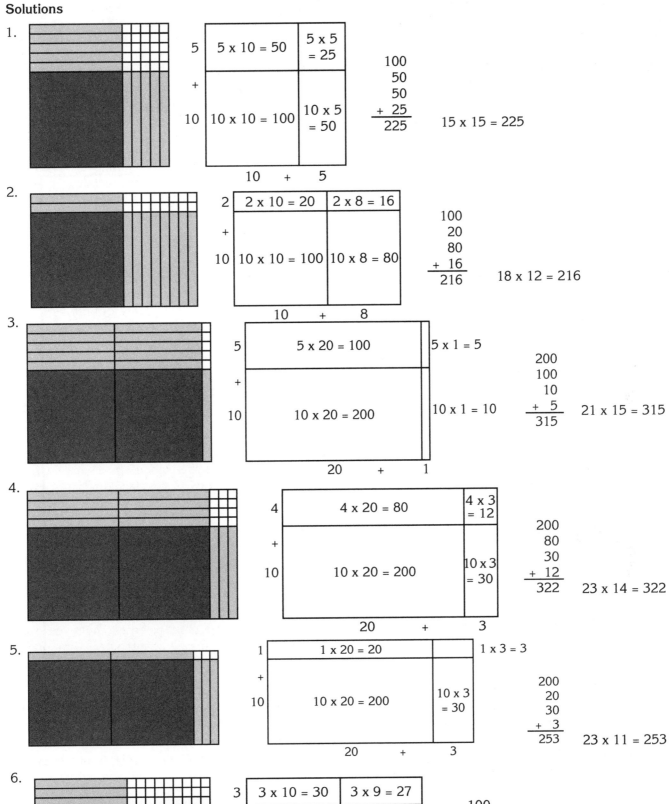

1.
5	5 x 10 = 50	5 x 5 = 25
+		
10	10 x 10 = 100	10 x 5 = 50

10 + 5

```
 100
  50
  50
+ 25
 225
```
15 x 15 = 225

2.
2	2 x 10 = 20	2 x 8 = 16
+		
10	10 x 10 = 100	10 x 8 = 80

10 + 8

```
 100
  20
  80
+ 16
 216
```
18 x 12 = 216

3.
5	5 x 20 = 100	5 x 1 = 5
+		
10	10 x 20 = 200	10 x 1 = 10

20 + 1

```
 200
 100
  10
+  5
 315
```
21 x 15 = 315

4.
4	4 x 20 = 80	4 x 3 = 12
+		
10	10 x 20 = 200	10 x 3 = 30

20 + 3

```
 200
  80
  30
+ 12
 322
```
23 x 14 = 322

5.
1	1 x 20 = 20	1 x 3 = 3
+		
10	10 x 20 = 200	10 x 3 = 30

20 + 3

```
 200
  20
  30
+  3
 253
```
23 x 11 = 253

6.
3	3 x 10 = 30	3 x 9 = 27
+		
10	10 x 10 = 100	10 x 9 = 90

10 + 9

```
 100
  30
  90
+ 27
 247
```
19 x 13 = 247

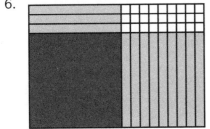

MULTIPLICATION THE ALGEBRA WAY

19

© 2012 AIMS Education Foundation

Building RECTANGLES

1.

X

Building **RECTANGLES**

2.

X

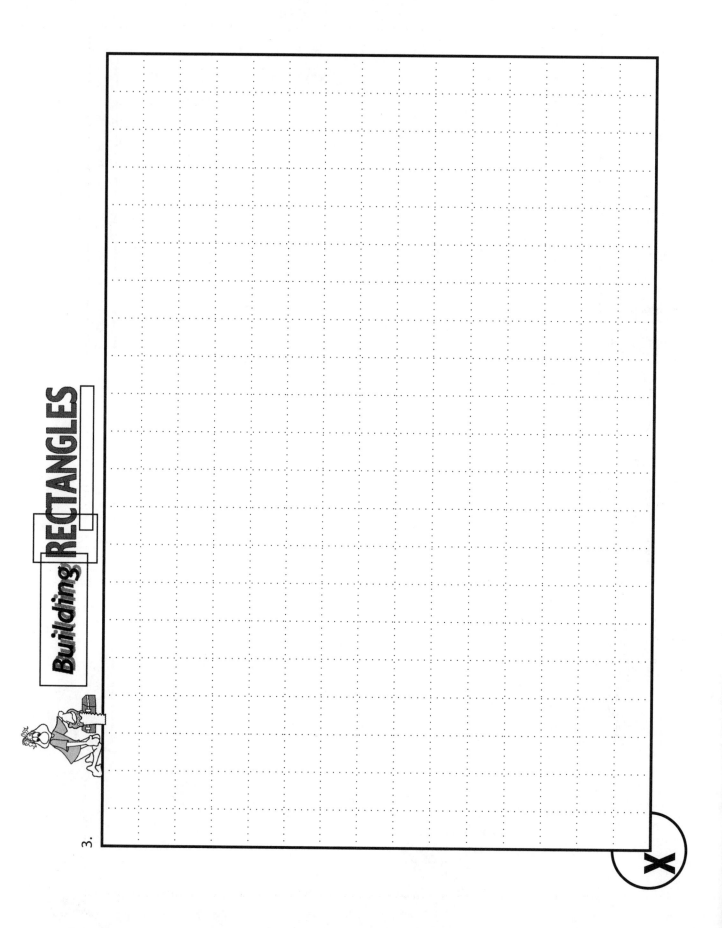

Building RECTANGLES

3.

X

Building RECTANGLES

4.

X

Building RECTANGLES

5.

X

6.

X

Connecting Learning

1. What strategies did you use to count up the number of cubes rapidly?

2. What patterns did you notice in how the types of pieces were placed in each rectangle?

3. What number patterns do you notice in the record of your solution?

4. How can you use your patterns to help you count the number of cubes in a rectangle if you are only told how long each side is?

Picturing RECTANGLES

Topic
Multiplication—representation

Key Question
How can you determine the number of squares in a rectangle if you are given the dimensions of the rectangle?

Learning Goals
Students will:
- learn that the solution of a multiplication problem is the sum of partial products, and
- learn to solve multi-digit multiplication problems with arrays.

Guiding Documents
Project 2061 Benchmarks
- *Add, subtract, multiply, and divide whole numbers mentally, on paper, and with a calculator.*
- *Express numbers like 100, 1000, and 1,000,000 as powers of 10.*

Common Core State Standards for Mathematics*
- *Reason abstractly and quantitatively. (MP.2)*
- *Use appropriate tools strategically. (MP.5)*
- *Look for and make use of structure. (MP.7)*
- *Use place value understanding and properties of operations to perform multi-digit arithmetic. (4.NBT.B)*

Math
Multiplication
Place value
Estimation

Integrated Processes
Observing
Comparing and contrasting
Generalizing

Materials
Base-ten blocks

Background Information
Students that have had opportunity to model two-digit multiplication with base-ten materials recognize that the resulting rectangle is divided into four regions.

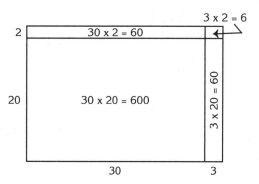

One region will be made of flats. Both of its dimensions will be multiples of ten.

Two regions will consist of sticks. One region will be above the flats with the sticks aligned horizontally. The second region will be to right side of the flats with the sticks aligned vertically. For both of these regions, one dimension will be a multiple of ten, and the other will be a unit digit.

The fourth region, diagonally opposite the flats, is made of unit cubes. The region's dimensions are both in units.

By asking students to construct the rectangle from the dimensions, this activity forces students to convert numerical information into a representational form. They may construct their representation either by recording it on a reduced grid; by making a box, line, and dot sketch; or by developing a rectangle marked into regions. Practice at the representational level will allow students to recognize the pattern of the regions' dimensions as decomposed place values of rectangles' dimensions, and solutions as the sum of the partial products of component regions. As students practice at the representational level, they will reinforce processes used at the abstract level.

Management
1. Prior experience is crucial in pacing this lesson. Students new to this process of multiplication may need to spend several days working at each level. Students with some experience may work through all the levels in one session mastering the concepts developed in *Building Rectangles*, *Picturing Rectangles*, and *Writing Rectangles*.

2. *Building Rectangles* should be done prior to this activity. *Building Rectangles* encourages students to recognize patterns in multiplication and moves them from a concrete model to a representation. This activity reinforces the patterns learned in *Building Rectangles* and encourages students to move from a representation to the utilization of the numerical equivalents.
3. Base-ten materials should be available for students to confirm their solutions gained using representational models.

Procedure

1. Present the *Key Question* to the class and discuss possible methods of solution.
2. Ask the students to picture a rectangle with dimensions of one of the samples below. Have them describe the shape of that rectangle.

27 x 13	42 x 21	25 x 13	43 x 11
33 x 22	52 x 13	32 x 23	61 x 12
32 x 27	67 x 11	33 x 31	76 x 11
32 x 32	85 x 11	32 x 31	94 x 11

3. Have the students use the dimensions to make a sketch of the rectangle. They may construct their representation either by recording it on a reduced grid; by making a box, line, and dot sketch; or by developing a rectangle marked into regions. If students are having difficulty visualizing the rectangle at this point, they may need more experience with building the rectangle using the base-ten pieces and then making a sketch of it.
4. Have the students use their sketches to count the number of squares in the rectangle and to share their counting methods.
5. Direct the students to record the dimensions of each region on the sketch. Also have them record the dimensions and areas of all four regions and find the sum of the parts.

$$30 \times 20 = 600$$
$$30 \times 2 = 60$$
$$3 \times 20 = 60$$
$$\underline{3 \times 2 = 6}$$
$$726$$

6. After students have sketched and solved a number of problems, discuss methods of getting a total number of squares within a rectangle without drawing the rectangle.

Connecting Learning

1. The rectangle for each problem divides into four regions. What pattern is there in how the regions divide each side's dimension? [The regions divide between the tens and the ones.]
2. What patterns did you notice in dimensions of each region? [The flats are always tens by tens, the sticks are always ones by tens, and the cubes are always ones by ones.]
3. How can you use your patterns to help you determine the number of squares in a rectangle if you are only told how long each side is? [Use the dimensions of the four regions to determine each partial product, then find the sum of the partial products.]

Extension

Have students use the patterns and processes they developed to solve a page of multiplication problems. Have them check their process using a calculator.

Picturing RECTANGLES

Use the grids below to make sketches of your rectangles.

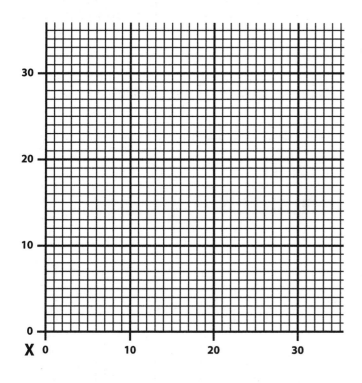

Record the dimensions and areas of all four regions. Find the sum of the parts.

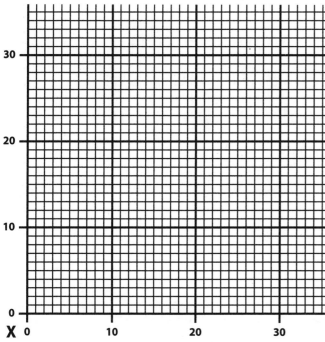

Picturing RECTANGLES

Connecting Learning

1. The rectangle for each problem divides into four regions. What pattern is there in how the regions divide each side's dimension?

2. What patterns did you notice in dimensions of each region?

3. How can you use your patterns to help you determine the number of squares in a rectangle if you are only told how long each side is?

MULTIPLICATION THE ALGEBRA WAY   30   © 2012 AIMS Education Foundation

Topic
Multiplication by partial products

Key Question
How can you determine the number of squares in a rectangle using only its dimensions?

Learning Goal
Students will learn to solve multi-digit multiplication problems with a partial product algorithm.

Guiding Documents
Project 2061 Benchmarks
- *Add, subtract, multiply, and divide whole numbers mentally, on paper, and with a calculator.*
- *Express numbers like 100, 1000, and 1,000,000 as powers of 10.*

Common Core State Standards for Mathematics*
- *Reason abstractly and quantitatively. (MP.2)*
- *Look for and make use of structure. (MP.7)*
- *Use place value understanding and properties of operations to perform multi-digit arithmetic. (4.NBT.B)*

Math
Multiplication
Place value
Estimation

Integrated Processes
Observing
Comparing and contrasting
Generalizing

Materials
Student record pages
Paper and pencil

Background Information
The partial product algorithm provides an excellent method for solving multi-digit multiplication problems. It provides a meaningful solution connected to concrete and representational models while retaining the place value meaning of the digits. It encourages students to practice multiplication by tens and develop their estimation capabilities.

The algorithm most often taught at the present time encourages students to ignore the place value of the digits and deal with them as digit facts to be placed in the appropriate column. In the problem

$$\begin{array}{r} 52 \\ \times\,34 \\ \hline \end{array}$$

students traditionally have been told to think of part of the solution as "3 x 2 and put the product of 6 directly under the 3." This provides the correct solution of six groups of 10 (60) but has distorted the meaning of the math to get a solution. The partial product method would have students view the same part as "30 groups of two," providing the corresponding product of 60. This method takes advantage of students' prior experience with base-ten materials. Using base-ten materials, the product of "3 x 2" would be one of the four regions in the 34 by 52 rectangle that was three sticks high and two squares wide using a total of six sticks and having a value of 60. By following this process, students maintain the full meaning of the place value system while practicing the important skill of estimation by powers of ten.

The movement of students from the representations of rectangular arrays to the abstraction of a multiplication algorithm is hastened by having the two methods demonstrated side by side. When the dimensions and products of each region are written in the representations, students are quick to see they are the same as the partial products produced by the algorithm.

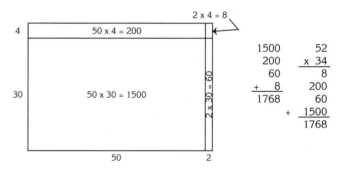

As students move to the algorithm, they utilize the image of the model in their heads. In the algorithm, they commonly do the tens by tens partial product first because it is easiest to count and takes up the largest area in the model. Although this is a reverse to most teacher's training, it can be embraced because it is meaningful and more appropriately provides a closer estimation of the final product.

As students spend time using the partial product algorithm, the listing of the factors begins to be done in their heads and only the products are recorded. This provides an efficient and meaningful way to approach multiplication.

Management
1. Prior experience is crucial in pacing this lesson. Students new to this method of multiplication may need to spend several days working at each level. Students with some experience may work through all the levels in one session mastering the concepts developed in *Building Rectangles, Picturing Rectangles*, and *Writing Rectangles*.
2. *Building Rectangles* and *Picturing Rectangles* should be done prior to this activity. *Writing Rectangles* encourages students to recognize patterns in multiplication and moves students from a representational model to an abstract algorithm.
3. Classes of students will vary in the amount of experience they will need to move from the representational level to an abstract level. Vary the amount of practice to meet the need of the students.

Procedure
1. Give the students the dimensions of the rectangle (31 x 24) and ask them to draw a representation of it and to record the dimensions along the edges, the squares in each region, and the total number of squares in the rectangle.
2. Model the partial product algorithm, and let the students try to determine why you are writing down each part.

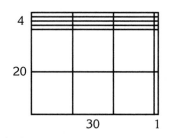

$$
\begin{array}{r}
31 \\
\times\ 24 \\
\hline
600 = 20 \times 30 \\
20 = 20 \times\ 1 \\
120 =\ \ 4 \times 30 \\
4 =\ \ 4 \times\ 1 \\
\hline
744
\end{array}
$$

3. Discuss the relationships between the sketches and the partial product algorithm.
4. When students understand the connections between the numerical and pictorial representations, have them complete the sketch for the sample problem and record the dimensions. Direct the students to follow your model of the partial product algorithm and compare it to the sketch.
5. When they have successfully completed several samples and are confident in the partial product concept, discuss how they might solve a multiplication problem in its vertical format. Encourage the students to justify their processes by referring to prior sketches and number processes.

6. Have them complete the multiplication problems in the vertical format using their partial product techniques and then check them with sketches.

Connecting Learning
1. What patterns do you notice in the dimensions of the rectangle and the dimensions of each region? [The dimensions in the regions use only one of the digits of each rectangle's dimension, either the tens or the ones.]
2. How is the teacher's method similar to your sketch of the rectangle? [There are four products and four regions; the four products use the same numbers as the regions' dimensions; the sum of the products is the same as the sum of the regions.]
3. How can you make sure you have all the correct dimensions for the regions if you only use the numbers? [Make sure all the digit values in one number are multiplied by all the digit values of the other number.]
4. How could you extend you number process to work with numbers with more than two digits? [Make sure you get all the regions by multiplying all the digit values in one number by all the digit values in the other.]

Extension
Have students try extending the algorithm to three digit problems and then check them with sketches or calculators.

Solutions

1.

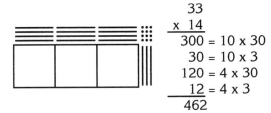

```
        33
      x 14
       300 = 10 x 30
        30 = 10 x 3
       120 = 4 x 30
        12 = 4 x 3
       462
```

2.

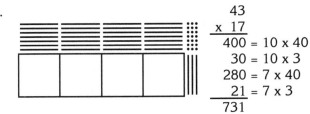

```
        32
      x 24
       600 = 20 x 30
        40 = 20 x 2
       120 = 4 x 30
         8 = 4 x 2
       768
```

3.

```
        43
      x 17
       400 = 10 x 40
        30 = 10 x 3
       280 = 7 x 40
        21 = 7 x 3
       731
```

4.

```
        42
      x 21
       800 = 20 x 40
        40 = 20 x 2
        40 = 1 x 40
         2 = 1 x 2
       882
```

5.

```
        54
      x 27
      1000 = 20 x 50
        80 = 20 x 4
       350 = 7 x 50
        28 = 7 x 4
      1458
```

6.

```
        27
      x 14
       200 = 10 x 20
        70 = 10 x 7
        80 = 4 x 20
        28 = 4 x 7
       378
```

	80	28
4		
+		
10	200	70
	20 +	7

7.

```
        35
      x 28
       600 = 20 x 30
       100 = 20 x 5
       240 = 8 x 30
        40 = 8 x 5
       980
```

	240	40
8		
+		
20	600	100
	30 +	5

8.

```
        42
      x 16
       400 = 10 x 40
        20 = 10 x 2
       240 = 6 x 40
        12 = 6 x 2
       672
```

	240	12
6		
+		
10	400	20
	40 +	2

9.

```
        55
      x 36
      1500 = 30 x 50
       150 = 30 x 5
       300 = 6 x 50
        30 = 6 x 5
      1980
```

	300	30
6		
+		
30	1500	150
	50 +	5

10.

```
        63
      x 34
      1800 = 30 x 60
        90 = 30 x 3
       240 = 4 x 60
        12 = 4 x 3
      2142
```

	240	12
4		
+		
30	1800	90
	60 +	3

Record

Make a sketch to determine the number of squares in each rectangle. Record the dimensions of each region, the squares in each region and the total squares in the rectangle.

Sketched Solution   **Number Solution**

1. length x width: 33 x 14

2. length x width: 32 x 24

3. length x width: 43 x 17

4. length x width: 42 x 21

5. length x width: 54 x 27

Writing RECTANGLES
Record

Use the partial product method to find the solution to each problem, then make a sketch to check your answer.

Number Solution   **Sketched Solution**

6. 27
    x 14
            = (x   )
            = (x   )
            = (x   )
            = (x   )
    _____

7. 35
    x 28
            = (x   )
            = (x   )
            = (x   )
            = (x   )
    _____

8. 42
    x 16
            = (x   )
            = (x   )
            = (x   )
            = (x   )
    _____

9. 55
    x 36
            = (x   )
            = (x   )
            = (x   )
            = (x   )
    _____

10. 63
    x 34
            = (x   )
            = (x   )
            = (x   )
            = (x   )
    _____

Writing RECTANGLES

Connecting Learning

1. What patterns do you notice in the dimensions of the rectangle and the dimensions of each region?

2. How is the teacher's method similar to your sketch of the rectangle?

3. How can you make sure you have all the correct dimensions for the regions if you only use the numbers?

4. How could you extend you number process to work with numbers with more than two digits?

Developing Understandings and Skills

The activities in *Part Two* build on and strengthen the understandings and skills students acquired in *Part One*. Duplicate sets of activities are provided making it possible to use one for an initial experience and the other for review at a later date; or one at one grade level and the other at a subsequent grade level.

Some students will be able to proceed through the sequence in this section rapidly while others will need more time and practice. It is important for each student to progress at a pace that produces optimum results.

Building mental images of algebraic processes and expressions is a major goal of these activities. The extensive use of manipulatives is helpful in attaining this objective.

A brief overview of the activities follows.

Display Multiplication

Students practice using the display (partial product) multiplication algorithm and thereby become increasingly aware of the "ten-ness" of our numeration system and the meaning this gives to the operation of multiplication. Because place value plays a central role, student understanding of this fundamental concept is nurtured. Two digits by two digits and two digits by three digits multiplication problems are included.

Expanding the View

In this activity, students are introduced to expanded notation, a staple in algebra. By keeping the vertical arrangement of multiplier and multiplicand, the relationship of this form to display multiplication as used in the previous activity is readily observable.

The use of expanded notation adds emphasis on the concept of place value. As students learn to write numbers in expanded notation, they cultivate their ability to think in terms of place value and the ten-ness of our numeration system.

Horizontal Multiplication

Taking the next step, students learn to write factors using expanded notation with a horizontal orientation, learn the sequence for finding partial products in correct order, and become increasingly aware of the role place value has in algebra. They practice the primary form of multiplication used in algebra while still in the familiar realm of the numbers of arithmetic.

Picturing Multiplication

Students learn to picture the multiplication of binomials and other polynomials, recognize that the pictured partial products are in one-to-one correspondence to those obtained by display multiplication, and build an understanding of an algorithm that will later help them think algebraically about multiplication of literal components.

Interpretations

Students learn to read and interpret pictures illustrating solutions to multiplication programs. They determine the factors and products involved and gain an understanding of what it means to use an inverse or "undoing" process. An understanding of doing and undoing is stressed in contemporary mathematics.

DISPLAY *Multiplication*

Topic
Multiplication—partial product

Key Question
How can multiplication be made conceptually rich?

Learning Goals
Students will:
- learn to multiply using the display multiplication method, and
- become increasingly aware of the "ten-ness" of our numeration system and the role it plays in multiplication.

Guiding Documents
Project 2061 Benchmark
- *Multiply whole numbers mentally and on paper.*

Common Core State Standards for Mathematics*
- *Reason abstractly and quantitatively. (MP.2)*
- *Look for and make use of structure. (MP.7)*
- *Use place value understanding and properties of operations to perform multi-digit arithmetic. (4.NBT.B)*

Math
Multiplication
Place value

Integrated Processes
Observing
Comparing and contrasting
Generalizing
Applying

Materials
Student sheets

Background Information
Language of Algebra
In display multiplication each sub-product is computed and recorded separately as shown in the example on the first student sheet. Students are also asked to write an expression indicating which digits were used in finding the product.

Notice the absence of splitting the digits in the product, "carrying" and "indenting." It is the straightforward way in which this process of multiplication is carried out that appeals to many students.

Often students ask, "Can we do it this way all the time?" The answer is yes, particularly if it is more meaningful. The question itself is a commentary on what we do too often in teaching mathematics: prescribe too narrowly how students can do things.

Concepts of Algebra
While the concept of place value is important in arithmetic, it takes on even greater significance in algebra as will be seen in the activities in this publication.

In display multiplication the emphasis is on the concept of place value (ten-ness of our numeration system) and the concept of multiplication. Students must constantly keep in mind the place value associated with each of the digits being multiplied.

Management
1. This activity has students multiply two- and three-digit numbers in a way that they have probably never seen before. You may want to go over a few examples as a class so that all students are clear on the procedure before they begin.
2. Students can work on this activity independently, or in small groups. There are advantages to both methods. You will have to decide which is better for your students.

Procedure
1. Hand out the student sheets and go over the instructions. Use the display method of multiplication to find the product of each of the problems given. You may want to do a few examples together as a class.
2. Have students work either alone or in small groups to complete both pages of problems.
3. When all students have finished, close with a time of class discussion and sharing.

Connecting Learning

1. How is this method different from regular multiplication? [It shows each of the sub-products.]
2. Did you find the display method easier or more difficult than standard multiplication? Why?
3. What did you learn about multiplication from this activity?
4. What did you learn about place value from this activity?

Extension

Have students try multiplying four- and five-digit numbers with the display method.

Solutions

Display Multiplication I

1.
```
      35
  x   13
      15 = ( 3 x   5)
      90 = ( 3 x  30)
      50 = (10 x   5)
  +  300 = (10 x  30)
     455
```

2.
```
      72
  x   24
       8 = ( 4 x   2)
     280 = ( 4 x  70)
      40 = (20 x   2)
  + 1400 = (20 x  70)
    1728
```

3.
```
      59
  x   36
      54 = ( 6 x   9)
     300 = ( 6 x  50)
     270 = (30 x   9)
  + 1500 = (30 x  50)
    2124
```

4.
```
      48
  x   27
      56 = ( 7 x   8)
     280 = ( 7 x  40)
     160 = (20 x   8)
  +  800 = (20 x  40)
    1296
```

5.
```
      87
  x   39
      63 = ( 9 x   7)
     720 = ( 9 x  80)
     210 = (30 x   7)
  + 2400 = (30 x  80)
    3393
```

6.
```
      63
  x   78
      24 = ( 8 x   3)
     480 = ( 8 x  60)
     210 = (70 x   3)
  + 4200 = (70 x  60)
    4914
```

7.
```
      92
  x   19
      18 = ( 9 x   2)
     810 = ( 9 x  90)
      20 = (10 x   2)
  +  900 = (10 x  90)
    1748
```

8.
```
      394
  x    76
       24 = ( 6 x     4)
      540 = ( 6 x    90)
     1800 = ( 6 x   300)
      280 = (70 x     4)
     6300 = (70 x    90)
  + 21000 = (70 x   300)
    29944
```

9.
```
      487
  x    65
       35 = ( 5 x     7)
      400 = ( 5 x    80)
     2000 = ( 5 x   400)
      420 = (60 x     7)
     4800 = (60 x    80)
  + 24000 = (60 x   400)
    31655
```

Display Multiplication II

1.
```
      42
  x   18
      16 = ( 8 x   2)
     320 = ( 8 x  40)
      20 = (10 x   2)
  +  400 = (10 x  40)
     756
```

2.
```
      75
  x   23
      15 = ( 3 x   5)
     210 = ( 3 x  70)
     100 = (20 x   5)
  + 1400 = (20 x  70)
    1725
```

3.
```
      91
  x   45
       5 = ( 5 x   1)
     450 = ( 5 x  90)
      40 = (40 x   1)
  + 3600 = (40 x  90)
    4095
```

4.
```
      38
  x   51
       8 = ( 1 x   8)
      30 = ( 1 x  30)
     400 = (50 x   8)
  + 1500 = (50 x  30)
    1938
```

5.
```
      84
  x   53
      12 = ( 3 x   4)
     240 = ( 3 x  80)
     200 = (50 x   4)
  + 4000 = (50 x  80)
    4452
```

6.
```
      67
  x   38
      56 = ( 8 x   7)
     480 = ( 8 x  60)
     210 = (30 x   7)
  + 1800 = (30 x  60)
    2546
```

7.
```
      68
  x   43
      24 = ( 3 x   8)
     180 = ( 3 x  60)
     320 = (40 x   8)
  + 2400 = (40 x  60)
    2924
```

8.
```
      256
  x    34
       24 = ( 4 x     6)
      200 = ( 4 x    50)
      800 = ( 4 x   200)
      180 = (30 x     6)
     1500 = (30 x    50)
  +  6000 = (30 x   200)
     8704
```

9.
```
      312
  x    98
       16 = ( 8 x     2)
       80 = ( 8 x    10)
     2400 = ( 8 x   300)
      180 = (90 x     2)
      900 = (90 x    10)
  + 27000 = (90 x   300)
    30576
```

DISPLAY Multiplication I

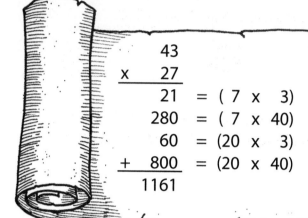

```
      43
   x  27
   ─────────
      21    = ( 7 x  3)
     280    = ( 7 x 40)
      60    = (20 x  3)
   +  800   = (20 x 40)
   ─────────
    1161
```

The example "displays" the sub-products created through multiplication. Please use this display method in finding the products in the following.

1.   35
   x 13
   ─────────
         ___ = (x   )
         ___ = (x   )
         ___ = (x   )
   + ___ ___ = (x   )

2.   72
   x 24
   ─────────
         ___ = (x   )
         ___ = (x   )
         ___ = (x   )
   + ___ ___ = (x   )

3.   59
   x 36
   ─────────
         ___ = (x   )
         ___ = (x   )
         ___ = (x   )
   + ___ ___ = (x   )

4.   48
   x 27
   ─────────
         ___ = (x   )
         ___ = (x   )
         ___ = (x   )
   + ___ ___ = (x   )

5.   87
   x 39
   ─────────
         ___ = (x   )
         ___ = (x   )
         ___ = (x   )
   + ___ ___ = (x   )

6.   63
   x 78
   ─────────
         ___ = (x   )
         ___ = (x   )
         ___ = (x   )
   + ___ ___ = (x   )

7.   92
   x 19
   ─────────
         ___ = (x   )
         ___ = (x   )
         ___ = (x   )
   + ___ ___ = (x   )

8.   394
   x 76
   ─────────
         ___ = (x   )
         ___ = (x   )
         ___ = (x   )
         ___ = (x   )
         ___ = (x   )
   + ___ ___ = (x   )

9.   487
   x 65
   ─────────
         ___ = (x   )
         ___ = (x   )
         ___ = (x   )
         ___ = (x   )
         ___ = (x   )
   + ___ ___ = (x   )

DISPLAY Multiplication II

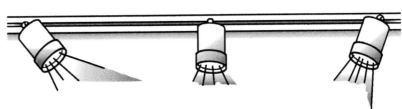

Using the same method you used in *Part One*, complete each of the problems below by writing the products in the appropriate spaces.

1.　42
　x　18

= (　 x 　)
= (　 x 　)
= (　 x 　)
+ ＿＿＿ = (　 x 　)

2.　75
　x　23

= (　 x 　)
= (　 x 　)
= (　 x 　)
+ ＿＿＿ = (　 x 　)

3.　91
　x　45

= (　 x 　)
= (　 x 　)
= (　 x 　)
+ ＿＿＿ = (　 x 　)

4.　38
　x　51

= (　 x 　)
= (　 x 　)
= (　 x 　)
+ ＿＿＿ = (　 x 　)

5.　84
　x　53

= (　 x 　)
= (　 x 　)
= (　 x 　)
+ ＿＿＿ = (　 x 　)

6.　67
　x　38

= (　 x 　)
= (　 x 　)
= (　 x 　)
+ ＿＿＿ = (　 x 　)

7.　68
　x　43

= (　 x 　)
= (　 x 　)
= (　 x 　)
+ ＿＿＿ = (　 x 　)

8.　256
　x　34

= (　 x 　)
= (　 x 　)
= (　 x 　)
= (　 x 　)
= (　 x 　)
+ ＿＿＿ = (　 x 　)

9.　312
　x　98

= (　 x 　)
= (　 x 　)
= (　 x 　)
= (　 x 　)
= (　 x 　)
+ ＿＿＿ = (　 x 　)

MULTIPLICATION THE ALGEBRA WAY

DISPLAY *Multiplication*

Connecting Learning

1. How is this method different from regular multiplication?

2. Did you find the display method easier or more difficult than standard multiplication? Why?

3. What did you learn about multiplication from this activity?

4. What did you learn about place value from this activity?

EXPANDING EXPANDING EXPANDING *the View*

Topic
Multiplication using expanded notation

Key Question
How does multiplication using expanded notation require thinking in terms of place value?

Learning Goals
Students will:
- learn how to write numbers in expanded notation, and
- become aware of the significance of thinking in terms of place value.

Guiding Documents
Project 2061 Benchmark
- *Multiply whole numbers mentally and on paper.*

Common Core State Standards for Mathematics*
- *Reason abstractly and quantitatively. (MP.2)*
- *Look for and make use of structure. (MP.7)*
- *Use place value understanding and properties of operations to perform multi-digit arithmetic. (4.NBT.B)*

Math
Multiplication
Place value

Integrated Processes
Observing
Comparing and contrasting
Generalizing
Applying

Materials
Student sheets

Background Information
Language of Algebra

In *Expanding the View*, students will compare and contrast display multiplication and multiplication using expanded notation. Both methods help to build understanding in preparation for multiplication in algebra. Expanded notation places an emphasis on place value. Ones, tens, hundreds, etc., are separated into individual terms. What has to be inferred when using normal notation is made explicit by such separation.

Concepts of Algebra

Mastery of the place value concept is essential if students are to build an understanding of the expressions and equations they will encounter in algebra. That is why so much emphasis is placed on introducing several methods of representation that increase the need to think in terms of place value.

Management
1. Be sure that students have completed the activity *Display Multiplication* before doing this activity, as they will be asked to use the display method and compare it to the expanded method.
2. You may want to do a few examples in addition to the one on the student sheet of how to use expanded display as a class before students begin so that they are comfortable with the procedure.

Procedure
1. Hand out the student sheets and go over the instructions, making sure that everyone understands the procedure. *Use both display and expanded display methods to compute the products on your student sheet.*
2. Have students work together in groups to complete the problems.
3. Close with a time of class discussion where students share their answers and what they learned from the process.

Connecting Learning
1. Which method did you find easier? Why?
2. What differences exist between the two methods? [In expanded notation you show the place value of the products.]
3. What did expanded display teach you about multiplication?
4. What did you learn about place value?

Extension
Have students try multiplying three-digit numbers using the expanded notation.

Solutions

Expanding the View I

1.
```
    47      40 + 7
  x 32    x 30 + 2
    14        14
    80        80
   210       210
 + 1200    + 1200
  1504      1504
```

2.
```
    43      40 + 3
  x 28    x 20 + 8
    24        24
   320       320
    60        60
 + 800     + 800
  1204      1204
```

3.
```
    54      50 + 4
  x 29    x 20 + 9
    36        36
   450       450
    80        80
 + 1000    + 1000
  1566      1566
```

4.
```
    39      30 + 9
  x 27    x 20 + 7
    63        63
   210       210
   180       180
 + 600     + 600
  1053      1053
```

5.
```
    63      60 + 3
  x 48    x 40 + 8
    24        24
   480       480
   120       120
 + 2400    + 2400
  3024      3024
```

6.
```
    56      50 + 6
  x 38    x 30 + 8
    48        48
   400       400
   180       180
 + 1500    + 1500
  2128      2128
```

7.
```
    77      70 + 7
  x 58    x 50 + 8
    56        56
   560       560
   350       350
 + 3500    + 3500
  4466      4466
```

Expanding the View II

1.
```
    52      50 + 2
  x 37    x 30 + 7
    14        14
   350       350
    60        60
 + 1500    + 1500
  1924      1924
```

2.
```
    46      40 + 6
  x 37    x 30 + 7
    42        42
   280       280
   180       180
 + 1200    + 1200
  1702      1702
```

3.
```
    61      60 + 1
  x 82    x 80 + 2
     2         2
   120       120
    80        80
 + 4800    + 4800
  5002      5002
```

4.
```
    93      90 + 3
  x 29    x 20 + 9
    27        27
   810       810
    60        60
 + 1800    + 1800
  2697      2697
```

5.
```
    35      30 + 5
  x 98    x 90 + 8
    40        40
   240       240
   450       450
 + 2700    + 2700
  3430      3430
```

6.
```
    87      80 + 7
  x 61    x 60 + 1
     7         7
    80        80
   420       420
 + 4800    + 4800
  5307      5307
```

7.
```
    74      70 + 4
  x 19    x 10 + 9
    36        36
   630       630
    40        40
 + 700     + 700
  1406      1406
```

8.
```
    17      10 + 7
  x 95    x 90 + 5
    35        35
    50        50
   630       630
 + 900     + 900
  1615      1615
```

* © Copyright 2010. National Governors Association Center for Best Practices and Council of Chief State School Officers. All rights reserved.

EXPANDING
EXPANDING
EXPANDING *the View 1*

Two methods of multiplication are shown in the example: display and expanded display. Please find the products using both of these methods and compare the processes. The order for computing by the second method is shown in the example.

Example:

```
    27        20 + 7
  x 14      x 10 + 4
  ────      ────────
    28          28
    80          80
    70          70
 + 200       + 200
 ─────       ──────
   378         378
```

1.   47   40 + 7
    x 32   x 30 + 2
    ────   ────────

2.   43
    x 28   _____

3.   54
    x 29   _____

4.   39
    x 27   _____

5.   63
    x 48   _____

6.   56
    x 38   _____

7.   77
    x 58   _____

Using the same process you used in *Part One*, find the products for each problem using both methods.

1. 52
   x 37   _____

2. 46
   x 37   _____

3. 61
   x 82   _____

4. 93
   x 29   _____

5. 35
   x 98   _____

6. 87
   x 61   _____

7. 74
   x 19   _____

8. 17
   x 95   _____

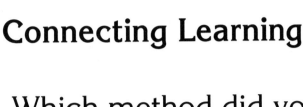

Connecting Learning

1. Which method did you find easier? Why?

2. What differences exist between the two methods?

3. What did expanded display teach you about multiplication?

4. What did you learn about place value?

HORIZONTAL *Multiplication*

Topic
Multiplication—distribution

Key Question
What is the algorithm for using the distributive property when factors are arranged horizontally?

Learning Goals
Students will:
- learn how to write factors in expanded notation for use in a horizontal arrangement,
- learn to use the algorithm for the distributive property in such an algebraic arrangement, and
- become increasingly aware of the significance of thinking in terms of place value.

Guiding Documents
Project 2061 Benchmark
- *Multiply whole numbers mentally and on paper.*

Common Core State Standards for Mathematics*
- *Reason abstractly and quantitatively. (MP.2)*
- *Look for and make use of structure. (MP.7*
- *Use place value understanding and properties of operations to perform multi-digit arithmetic. (4.NBT.B)*

Math
Multiplication
Place value

Integrated Processes
Observing
Comparing and contrasting
Generalizing
Applying

Materials
Student sheets

Background Information
Language of Algebra

In *Horizontal Multiplication*, students will compare and contrast display multiplication and the general algebraic method of multiplication. For this, they must learn the language specific to this algorithm as shown in this example: 34 x 47 =

First, both numbers are written in expanded notation: $(30 + 4)(40 + 7) =$ Students need to learn that parentheses, such as those above, indicate multiplication in algebra.

Next, both numbers in the second factor are multiplied by the first number in the first factor, and then by the second number in the first factor:

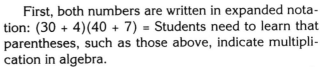

$(30 + 4) (40 + 7) =$

$(30 \times 40) + (30 \times 7) + (4 \times 40) + (4 \times 7) =$
$\quad\; 1 \qquad\qquad 2 \qquad\qquad 3 \qquad\qquad 4$

Finally, the four sub-products are added: 1200 + 210 + 160 + 28 = 1598

As students compare the sub-products resulting from the two methods of multiplication, they will see that they are the same but arranged in opposite order.

Concepts of Algebra

As in previous activities, emphasis remains on the concepts of place value and multiplication. The focus here is particularly on the use of the distributive property.

Management
1. Students need to have completed the activity *Display Multiplication* before attempting this one. *Horizontal Multiplication* asks students to use both the display method and the distributive property method to solve multiplication problems. If students need a review of the display method, do a few problems as a class before giving them the activity.
2. If students are unclear on the distributive property method, you can do several examples as a class in addition to the one provided on the student sheet.

Procedure
1. Hand out the student sheets and go over the instructions. *Solve each problem using both the display and the distributive property methods. Compare the processes and the sub-products.*
2. Have students work together in small groups to complete both student sheets.
3. Close the activity with a time of class discussion and sharing.

Connecting Learning

1. How is the distributive property method different from the display method?
2. What similarities do they have?
3. What do you notice about the sub-products in the two different methods? [They are the same, but appear in the opposite order.]
4. Which method did you find easier? Why?
5. What did you learn about multiplication from this activity?

Extension

Have students try the distributive property method with three and/or four digit numbers.

Solutions

Horizontal Multiplication Part One

1.
```
     39
   x 64
   ----
     36
    120
    540
  +1800
   ----
   2496
```
$(30 + 9)(60 + 4) =$

$(30 \times 60) + (30 \times 4) +$
$(9 \times 60) + (9 \times 4) =$

$1800 + 120 + 540 + 36 = 2496$

2.
```
     42
   x 97
   ----
     14
    280
    180
  +3600
   ----
   4074
```
$(40 + 2)(90 + 7) =$

$(40 \times 90) + (40 \times 7) +$
$(2 \times 90) \quad + (2 \times 7) =$

$3600 + 280 + 180 + 14 = 4074$

3.
```
     65
   x 56
   ----
     30
    360
    250
  +3000
   ----
   3640
```
$(60 + 5)(50 + 6) =$

$(60 \times 50) + (60 \times 6) +$
$(5 \times 50) + (5 \times 6) =$

$3000 + 360 + 250 + 30 = 3640$

4.
```
     89
   x 78
   ----
     72
    640
    630
  +5600
   ----
   6942
```
$(80 + 9)(70 + 8) =$

$(80 \times 70) + (80 \times 8) +$
$(9 \times 70) + (9 \times 8) =$

$5600 + 640 + 630 + 72 = 6942$

5.
```
     38
   x 43
   ----
     24
     90
    320
  +1200
   ----
   1634
```
$(30 + 8)(40 + 3) =$

$(30 \times 40) + (30 \times 3) +$
$(8 \times 40) + (8 \times 3) =$

$1200 + 90 + 320 + 24 = 1634$

6.
```
     57
   x 96
   ----
     42
    300
    630
  +4500
   ----
   5472
```
$(50 + 7)(90 + 6) =$

$(50 \times 90) + (50 \times 6) +$
$(7 \times 90) + (7 \times 6) =$

$4500 + 300 + 630 + 42 = 5472$

1. $\begin{array}{r} 36 \\ \times\ 44 \\ \hline 24 \\ 120 \\ 240 \\ +\ 1200 \\ \hline 1584 \end{array}$

    (30 + 6)(40 + 4) =

    (30 x 40) + (30 x 4) +
    (6 x 40) + (6 x 4) =

    1200 + 120 + 240 + 24 = 1584

2. $\begin{array}{r} 28 \\ \times\ 52 \\ \hline 16 \\ 40 \\ 400 \\ +\ 1000 \\ \hline 1456 \end{array}$

    (20 + 8)(50 + 2) =

    (20 x 50) + (20 x 2) +
    (8 x 50) + (8 x 2) =

    1000 + 40 + 400 + 16 = 1456

3. $\begin{array}{r} 27 \\ \times\ 62 \\ \hline 14 \\ 40 \\ 420 \\ +\ 1200 \\ \hline 1674 \end{array}$

    (20 + 7)(60 + 2) =

    (20 x 60) + (20 x 2) +
    (7 x 60) + (7 x 2) =

    1200 + 40 + 420 + 14 = 1674

4. $\begin{array}{r} 72 \\ \times\ 38 \\ \hline 16 \\ 560 \\ 60 \\ +\ 2100 \\ \hline 2736 \end{array}$

    (70 + 2)(30 + 8) =

    (70 x 30) + (70 x 8) +
    (2 x 30) + (2 x 8) =

    2100 + 560 + 60 + 16 = 2736

5. $\begin{array}{r} 13 \\ \times\ 87 \\ \hline 21 \\ 70 \\ 240 \\ +\ 800 \\ \hline 1131 \end{array}$

    (10 + 3)(80 + 7) =

    (10 x 80) + (10 x 7) +
    (3 x 80) + (3 x 7) =

    800 + 70 + 240 + 21 = 1131

6. $\begin{array}{r} 22 \\ \times\ 94 \\ \hline 8 \\ 80 \\ 180 \\ +\ 1800 \\ \hline 2068 \end{array}$

    (20 + 2)(90 + 4) =

    (20 x 90) + (20 x 4) +
    (2 x 90) + (2 x 4) =

    1800 + 80 + 180 + 8 = 2068

7. $\begin{array}{r} 41 \\ \times\ 73 \\ \hline 3 \\ 120 \\ 70 \\ +\ 2800 \\ \hline 2993 \end{array}$

    (40 + 1)(70 + 3) =

    (40 x 70) + (40 x 3) +
    (1 x 70) + (1 x 3) =

    2800 + 120 + 70 + 3 = 2993

8. $\begin{array}{r} 85 \\ \times\ 14 \\ \hline 20 \\ 320 \\ 50 \\ +\ 800 \\ \hline 1190 \end{array}$

    (80 + 5)(10 + 4) =

    (80 x 10) + (80 x 4) +
    (5 x 10) + (5 x 4) =

    800 + 320 + 50 + 20 = 1190

HOR-NOZIHAL Multiplication

Part One

The algebraic use of the distributive property of multiplication over addition is shown here in its numeric form together with the display method. Please solve each problem using both methods. Compare the processes and sub-products.

Display Method:

$$27$$
$$\underline{\times 38}$$
$$56$$
$$160$$
$$210$$
$$\underline{+ 600}$$
$$1026$$

Distributive Property Method:

$(20 + 7)(30 + 8) =$

$(20 \times 30) + (20 \times 8) + (7 \times 30) + (7 \times 8) =$

$600 + 160 + 210 + 56 = 1026$

1. 39
    $\underline{\times\ 64}$ $(30 + 9)(60 + 4) =$

2. 42
    $\underline{\times\ 97}$ $(\quad + \quad)(\quad + \quad) =$

3. 65
    $\underline{\times\ 56}$ $(\quad + \quad)(\quad + \quad) =$

4. 89
    $\underline{\times\ 78}$ $(\quad + \quad)(\quad + \quad) =$

5. 38
    $\underline{\times\ 43}$ $(\quad + \quad)(\quad + \quad) =$

6. 57
    $\underline{\times\ 96}$ $(\quad + \quad)(\quad + \quad) =$

HORIZONTAL Multiplication

Part Two

Solve each problem below using the distributive property and the display method as you did in *Part One*.

1. 36 $(30 + 6)(40 + 4) =$
    x 44

2. 28 $(\quad + \quad)(\quad + \quad) =$
    x 52

3. 27 $(\quad + \quad)(\quad + \quad) =$
    x 62

4. 72 $(\quad + \quad)(\quad + \quad) =$
    x 38

5. 13 $(\quad + \quad)(\quad + \quad) =$
    x 87

6. 22 $(\quad + \quad)(\quad + \quad) =$
    x 94

7. 41 $(\quad + \quad)(\quad + \quad) =$
    x 73

8. 85 $(\quad + \quad)(\quad + \quad) =$
    x 14

HOR–IZON–TAL *Multiplication*

Connecting Learning

1. How is the distributive property method different from the display method?

2. What similarities do they have?

3. What do you notice about the sub-products in the two different methods?

4. Which method did you find easier? Why?

5. What did you learn about multiplication from this activity?

Picturing Multiplication

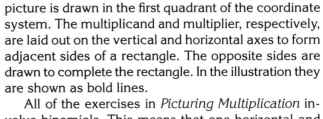

Multiplication

Topic
Multiplication—representation

Key Question
How can the multiplication of binomials be pictured?

Learning Goals
Students will:
- learn how to picture multiplication of binomials and other polynomials,
- recognize that the pictured subproducts are the same as those obtained by the display and other methods of multiplication, and
- build a foundation of understanding that will later help them think algebraically about multiplication with literal components.

Guiding Documents
Project 2061 Benchmark
- *Multiply whole numbers mentally and on paper.*

Common Core State Standards for Mathematics*
- *Reason abstractly and quantitatively. (MP.2)*
- *Use appropriate tools strategically. (MP.5)*
- *Look for and make use of structure. (MP.7*
- *Use place value understanding and properties of operations to perform multi-digit arithmetic. (4.NBT.B)*

Math
Multiplication
Place value

Integrated Processes
Observing
Comparing and contrasting
Generalizing
Applying

Materials
Student sheets
Rulers

Background Information
Language of Algebra
In *Picturing Multiplication*, students solve each problem in three ways: using the display method, horizontal multiplication, and by picturing the result. What needs to be introduced is the standard algebraic graphical representation of multiplication using positive numbers.

Since the product in these activities is positive, the picture is drawn in the first quadrant of the coordinate system. The multiplicand and multiplier, respectively, are laid out on the vertical and horizontal axes to form adjacent sides of a rectangle. The opposite sides are drawn to complete the rectangle. In the illustration they are shown as bold lines.

All of the exercises in *Picturing Multiplication* involve binomials. This means that one horizontal and one vertical line will be drawn through the rectangle from each point dividing tens and ones. This creates four smaller rectangles.

If all the horizontal and vertical unit lines were to be drawn through the rectangle, they would create the same number of small unit rectangles in each smaller rectangle outlined in bold as the number in the respective subproducts.

This shows the multiplication of 23 x 15:

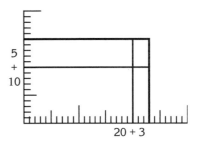

Concepts of Algebra
These activities continue the development of the concepts of multiplication and place value. By using multiple representations, students have the opportunity of viewing a given situation from multiple perspectives. This leads to greater depth of understanding.

Management
1. Students must be familiar with the display and expanded display methods of multiplication before beginning this activity. If students need to be reminded of these procedures, be sure to have a review before beginning the activity.
2. Rulers should be used by students to make neat and accurate rectangles in the grids.

Procedure
1. Hand out the student sheets and go over the instructions. *Use three methods to solve each problem: display, expanded display, and graphic notation.*
2. Have students work together in small groups to complete the student sheets.
3. Close the activity with a time of class discussion and sharing.

Connecting Learning

1. How does showing the products graphically compare to the first two methods?
2. What did this representation show you about the products?
3. How are the subproducts related to the small rectangles created on the grids? [The number of unit rectangles in each small rectangle is equal to the value of the subproduct represented by that rectangle.]
4. What did using the graphic method teach you about multiplication?

Extension

Have students create their own multiplication problems to represent graphically.

Solutions

For the solutions, only the graphs of the problems are shown.

Picturing Multiplication I

1.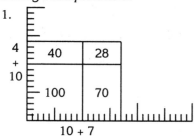

2.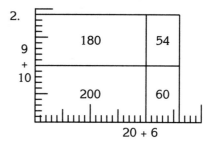

3.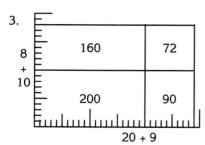

4.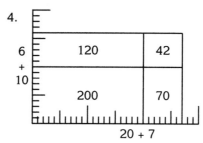

5.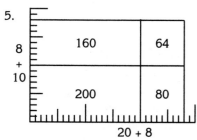

6.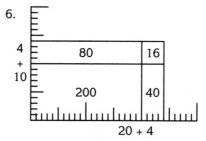

7.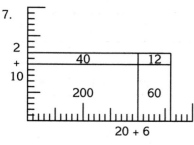

8.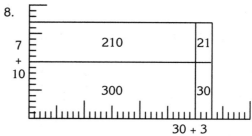

1.
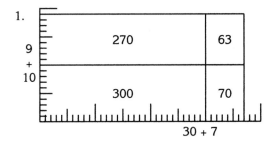

2.
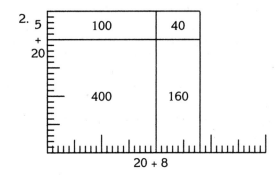

3.
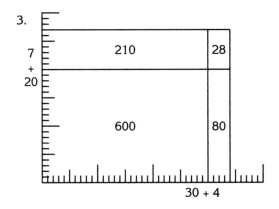

4.
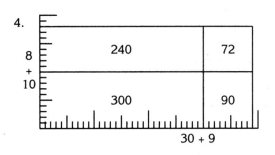

5.
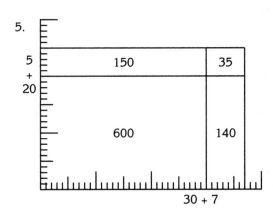

6.
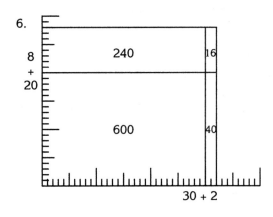

7.
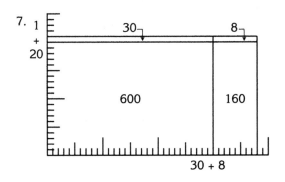

8.
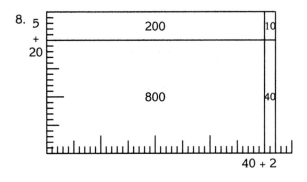

Picturing Multiplication I

Please show multiplication in three ways using the (a) display and (b) expanded notation algorithms, and (c) graphically by sketching a four-section rectangle showing dimensions and sub-products. Write each subproduct in the appropriate section of the rectangle in the grid. Compare the three results. The first problem has been done for you as an example.

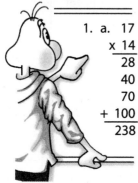

1. a.   17
        x 14

          28
          40
          70
       + 100

         238

    b. $(10 + 7)(10 + 4) = (10 \cdot 10) + (10 \cdot 4) + (7 \cdot 10) + (7 \cdot 4)$
    $= 100 + 40 + 70 + 28 = 238$

    c.
    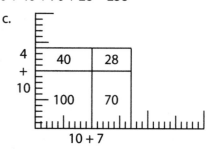

2. a.   26
        x 19

    b. $(\quad + \quad)(\quad + \quad) = (\quad \cdot \quad) + (\quad \cdot \quad) + (\quad \cdot \quad) + (\quad \cdot \quad)$
    =

    c.
    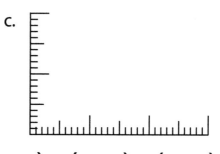

3. a.   29
        x 18

    b. $(\quad + \quad)(\quad + \quad) = (\quad \cdot \quad) + (\quad \cdot \quad) + (\quad \cdot \quad) + (\quad \cdot \quad)$
    =

    c.
    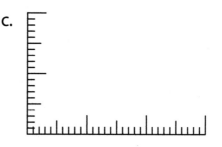

4. a.   27
        x 16

    b. $(\quad + \quad)(\quad + \quad) = (\quad \cdot \quad) + (\quad \cdot \quad) + (\quad \cdot \quad) + (\quad \cdot \quad)$
    =

    c.
    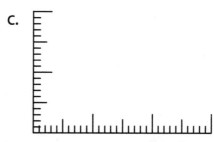

Picturing Multiplication I

Multiplication

5. a. 28
   x 18

   b. (+) (+) = (·) + (·) + (·) + (·)
      =

   c.

6. a. 24
   x 14

   b. (+) (+) = (·) + (·) + (·) + (·)
      =

   c.

7. a. 26
   x 12

   b. (+) (+) = (·) + (·) + (·) + (·)
      =

   c.

8. a. 33
   x 17

   b. (+) (+) = (·) + (·) + (·) + (·)
      =

   c.

Picturing Multiplication II

Please show multiplication in three ways using the (a) display and (b) expanded notation algorithms, and (c) graphically by sketching a four-section rectangle showing dimensions and sub-products. Compare the three results. Write each subproduct in the appropriate section of the rectangle in the grid.

1. a. 37
   x 19

   b. (+   ) (+   ) = (·   ) + (·   ) + (·   ) + (·   )
   =

   c.

2. a. 28
   x 25

   b. (+   ) (+   ) = (·   ) + (·   ) + (·   ) + (·   )
   =

   c.

3. a. 34
   x 27

   b. (+   ) (+   ) = (·   ) + (·   ) + (·   ) + (·   )
   =

   c.

4. a. 39
   x 18

   b. (+   ) (+   ) = (·   ) + (·   ) + (·   ) + (·   )
   =

   c.

Picturing Multiplication II

Multiplication

5. a. 37   b. (+   ) (+   ) = (·   ) + (·   ) + (·   ) + (·   )

     x 25   =   c.

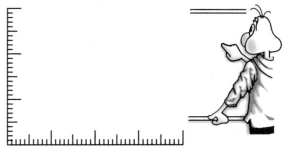

6. a. 32   b. (+   ) (+   ) = (·   ) + (·   ) + (·   ) + (·   )

     x 28   =

                                      c.

7. a. 38   b. (+   ) (+   ) = (·   ) + (·   ) + (·   ) + (·   )

     x 21   =

                                      c.

8. a. 42   b. (+   ) (+   ) = (·   ) + (·   ) + (·   ) + (·   )

     x 25   =

                                      c.

Picturing Multiplication

Multiplication

Connecting Learning

1. How does showing the products graphically compare to the first two methods?

2. What did this representation show you about the products?

3. How are the subproducts related to the small rectangles created on the grids?

4. What did using the graphic method teach you about multiplication?

Interpretations

Topic
Multiplication—representation

Key Question
How can we determine factors and products in multiplication from pictures?

Learning Goals
Students will:
- learn to read and interpret pictures resulting from multiplication to determine the factors and products involved, and
- gain understanding of what it means to use an inverse or "undoing" process.

Guiding Documents
Project 2061 Benchmark
- *Multiply whole numbers mentally and on paper.*

Common Core State Standards for Mathematics*
- *Reason abstractly and quantitatively. (MP.2)*
- *Use appropriate tools strategically. (MP.5)*
- *Look for and make use of structure. (MP.7)*
- *Use place value understanding and properties of operations to perform multi-digit arithmetic. (4.NBT.B)*

Math
Multiplication
Place value

Integrated Processes
Observing
Comparing and contrasting
Generalizing
Applying

Materials
Student sheets
Rulers

Background Information
Language of Algebra

In *Interpretations* students determine the factors and products involved in a multiplication by reading pictures. This is the inverse of the process used in *Picturing Multiplication*. They are asked to record the factors in expanded notation and the product in standard notation.

Concepts of Algebra

The *Picturing Multiplication* and *Interpretations* activities exemplify the "doing-undoing" processes in algebra and begin to orient student thinking to the use of inverse approaches. The content of algebra is replete with "doing-undoing" usage. It is important that this becomes a way of thinking with students, a habit of the mind as they grow in their ability to engage in algebraic thinking.

Management
1. This activity should be done following *Picturing Multiplication* so that students can compare the differing procedures that yield the same result.
2. Students should use rulers to make the lines as neat as possible.

Procedure
1. Hand out the student sheets and go over the instructions. *Divide the area within each outline into smaller regions to show hundreds, tens, and ones. Write the mathematical sentence length x width = area for each picture using expanded notation.*
2. Have students work together in small groups to complete the student sheets.
3. Close with a time of class discussion and sharing.

Connecting Learning
1. How is this process like the one you used in *Picturing Multiplication*? [You are dividing a rectangular area into hundreds, tens, and ones, showing each subproduct graphically.]
2. How is it different? [You don't have the values of the numbers to begin with; you determine them based on the size of the rectangle.]
3. What do the smaller regions within the rectangle represent? [The value of the subproducts in the equation.]
4. What did you learn about the nature of multiplication from this activity?

Extension
Have students make their own problems and trade them with classmates.

Solutions

Interpretations I

1.

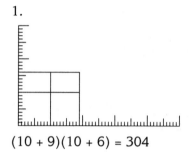

(10 + 9)(10 + 6) = 304

2.

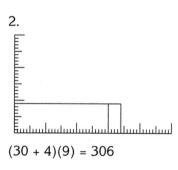

(30 + 4)(9) = 306

3.

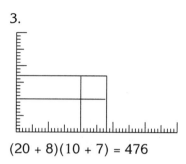

(20 + 8)(10 + 7) = 476

4.

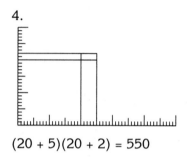

(20 + 5)(20 + 2) = 550

5.

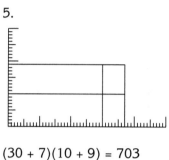

(30 + 7)(10 + 9) = 703

6.

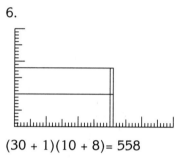

(30 + 1)(10 + 8)= 558

7.

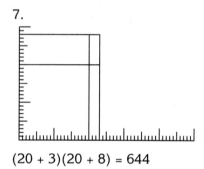

(20 + 3)(20 + 8) = 644

8.

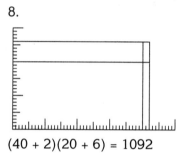

(40 + 2)(20 + 6) = 1092

9.

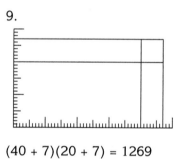

(40 + 7)(20 + 7) = 1269

10.

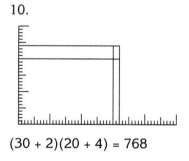

(30 + 2)(20 + 4) = 768

11.

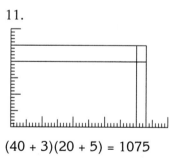

(40 + 3)(20 + 5) = 1075

12.

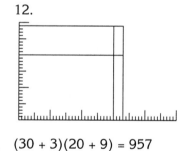

(30 + 3)(20 + 9) = 957

1.

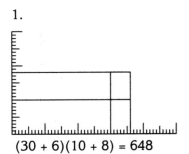

(30 + 6)(10 + 8) = 648

2.

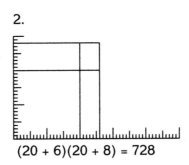

(20 + 6)(20 + 8) = 728

3.

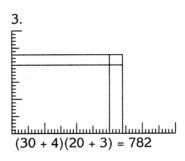

(30 + 4)(20 + 3) = 782

4.

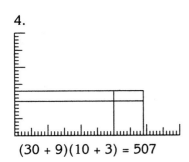

(30 + 9)(10 + 3) = 507

5.

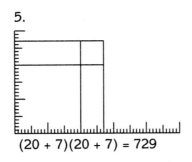

(20 + 7)(20 + 7) = 729

6.

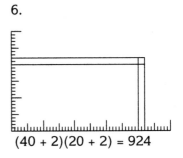

(40 + 2)(20 + 2) = 924

7.

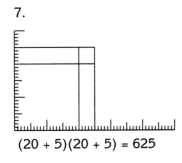

(20 + 5)(20 + 5) = 625

8.

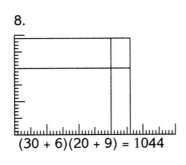

(30 + 6)(20 + 9) = 1044

9.

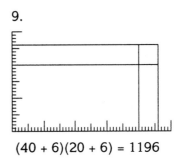

(40 + 6)(20 + 6) = 1196

10.

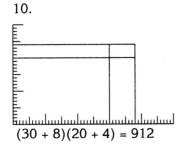

(30 + 8)(20 + 4) = 912

11.

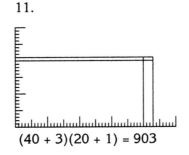

(40 + 3)(20 + 1) = 903

12.

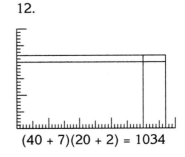

(40 + 7)(20 + 2) = 1034

Interpretations I

Please divide the area within each outline into smaller regions to show hudreds, tens, and ones. Write the mathematical sentences length x width = area for each picture using expanded notation. For example, 58 would be written as 50 + 8, and 256 as 200 + 50 + 6. Carry out the indicated multiplication. The first problem has been done for you as an example.

1.

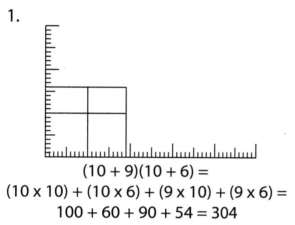

$$(10 + 9)(10 + 6) =$$
$$(10 \times 10) + (10 \times 6) + (9 \times 10) + (9 \times 6) =$$
$$100 + 60 + 90 + 54 = 304$$

2.

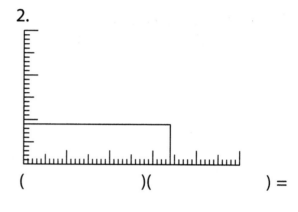

(   )(   ) =

3.

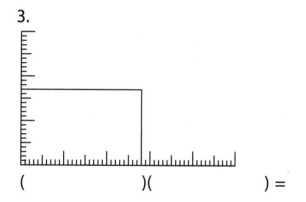

(   )(   ) =

4.

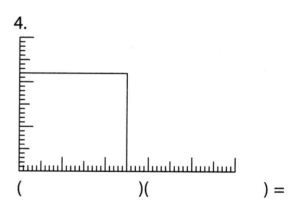

(   )(   ) =

5.

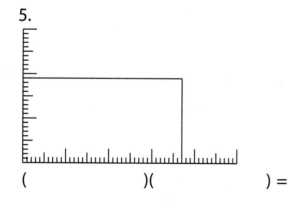

(   )(   ) =

6.

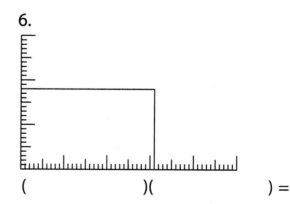

(   )(   ) =

Interpretations I

7.

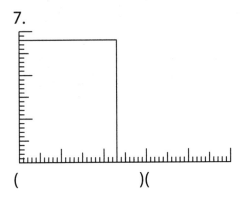

(   )(   ) =

8.

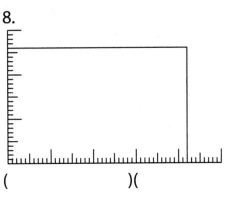

(   )(   ) =

9.

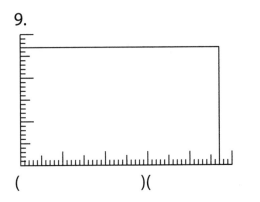

(   )(   ) =

10.

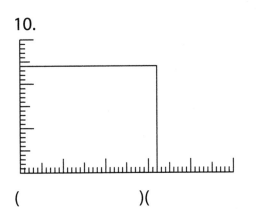

(   )(   ) =

11.

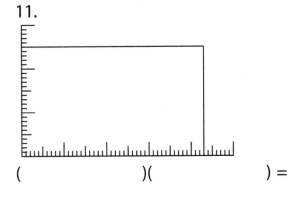

(   )(   ) =

12.

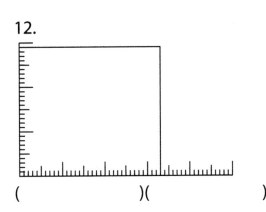

(   )(   ) =

Interpretations II

Use the same method you used in *Part One* to complete the following problems using expanded and graphic notation.

1.

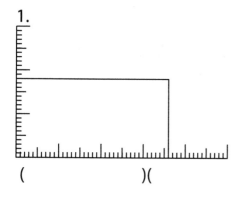

(   )(   ) =

2.

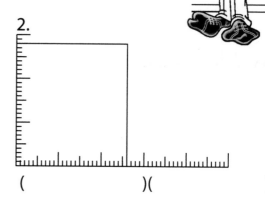

(   )(   ) =

3.

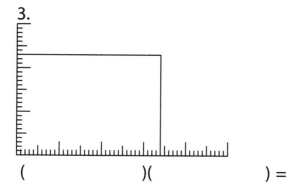

(   )(   ) =

4.

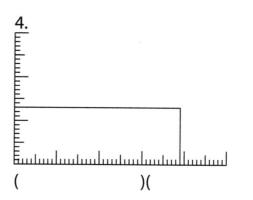

(   )(   ) =

5.

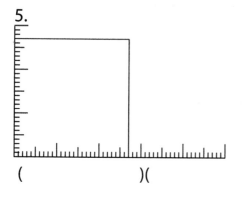

(   )(   ) =

6.

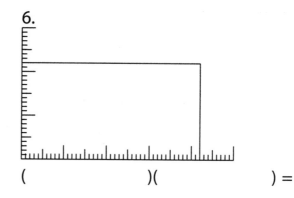

(   )(   ) =

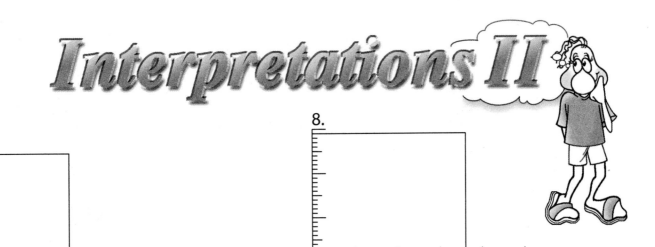

7.

( )( ) =

8.

( )( ) =

9.

( )( ) =

10.

( )( ) =

11.

( )( ) =

12.

( )( ) =

Interpretations

Connecting Learning

1. How is this process like the one you used in *Picturing Multiplication*?

2. How is it different?

3. What do the smaller regions within the rectangle represent?

4. What did you learn about the nature of multiplication from this activity?

Introducing Quadratic Expressions and Equations in Base Ten

After students have learned the prerequisite skills and understandings, they are ready to study quadratic expressions and equations. The concepts that occur in this series of activities will later be transferred to other number bases and algebraic expressions. Learning the concepts in the familiar context of arithmetic numbers makes them more meaningful. This will result in better understanding as they apply the concepts in other number bases and use the literal notation of algebra.

Filling Frames

Students fill in rectangles in the first quadrant using the fewest possible number of base-ten blocks. From the result, they determine the factors—the length, and width, and area—that are their product. They may be surprised that each rectangle has a unique solution. The process models factoring in algebra. The product is a quadratic expression in base ten and it is factored into a unique set of factors.

Models of Square Numbers

Students fill outlines of frames representing consecutive square numbers with the fewest possible number of base-ten blocks, determine the number being squared, and express its square in expanded notation. From the pattern that emerges, they should be able to develop a process for mentally computing the squares of numbers.

Constructions Plus

Given a defined set of base-ten blocks, students form them into a rectangle in the first quadrant following a prescribed procedure. The number of each type of block is the coefficient of a term in a quadratic expression. For example, if the set of blocks consists of 3 flats, 11 longs, and 6 units, the base-ten expression is $3(10^2) + 11(10) + 6(10^0)$. The equivalent algebraic expression is $3x^2 + 11x + 6$. The factors are $3(10) + 2$ and $10 + 3$ or $(3x + 2)(x + 3)$ where $x = 10$. The quadratic equation under consideration is $(3x + 2)(x + 3) = 3x^2 + 11x + 6$.

Topic
Distributive property

Key Question
How can you find the length, width, and area of a rectangle using base-ten blocks?

Learning Goals
Students will:
- cover rectangles using the fewest possible number of base-ten blocks,
- be able to write a mathematical sentence for length x width = area using expanded notation, and
- associate each region formed by the blocks with its parallel in the mathematical sentence.

Guiding Document
Common Core State Standards for Mathematics*
- *Model with mathematics. (MP.4)*
- *Use appropriate tools strategically. (MP.5)*
- *Look for and make use of structure. (MP.7)*
- *Use place value understanding and properties of operations to perform multi-digit arithmetic. (4.NBT.B)*

Math
Distributive property
Multiplication
Expanded notation
Base ten

Integrated Processes
Observing
Collecting and
   recording data
Generalizing

Materials
Base-ten blocks
Rulers, optional
Student sheets

Background Information
To build a deep understanding of the distributive property in all of its uses in arithmetic and algebra, it is necessary to build associations between manipulatives, representations, and abstract expressions. In this activity students will manipulate base-ten blocks, draw pictures of their solutions, and interpret the pictures into an abstract expression using expanded notation.

Meaning is developed by associating each of the four regions in the picture with its abstract expression. It is this aspect that should receive principal emphasis. If students are to understand algebra and find it meaningful, they must be able to visualize algebraic expressions. This activity helps achieve such understanding.

Management
1. Students will each need base-ten blocks to complete this activity. One set can be shared between two or three students.
2. Students should be familiar with how to use base-ten blocks and how to write an equation in expanded notation before beginning this activity.
3. There are two parts to the student sheets: *Filling Frames I* and *Filling Frames II*. The second portion is for practice, and is not necessary if students have a good handle on the concept after the first half.
4. Students may want rulers to help them draw their sketches more neatly and accurately.

Procedure
1. Hand out the first three student sheets and base-ten blocks to each student and go over the instructions. *Cover each of the rectangles with base-ten blocks and write a mathematical sentence indicating the area using expanded notation.*
2. Have students work individually or in small groups to complete the student sheet. Be sure that they understand how to determine which region of their sketch corresponds to which part of the equation.
3. When students have completed the first section you may hand out the second section, for more practice or move on to a time of class discussion and sharing.

Connecting Learning
1. What did you discover about the way that the base-ten blocks are arranged in each rectangle? [There are four regions: hundreds, two groups of tens, and a group of ones.]
2. How are these regions related to the expanded notation version of the multiplication problem? [Each region represents one of the numbers in expanded notation.]
3. How does this relationship help you to understand the meaning of the physical representation?
4. Was this method of looking at multiplication easier or harder for you than other methods you have done previously? Explain.

Solutions

The solutions for both parts of this activity are included; however, the diagrams are only included for the solutions to *Filling Frames 1*.

Filling Frames 1

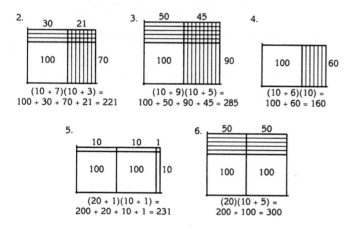

2.
$(10 + 7)(10 + 3) =$
$100 + 30 + 70 + 21 = 221$

3.
$(10 + 9)(10 + 5) =$
$100 + 50 + 90 + 45 = 285$

4.
$(10 + 6)(10) =$
$100 + 60 = 160$

5.
$(20 + 1)(10 + 1) =$
$200 + 20 + 10 + 1 = 231$

6.
$(20)(10 + 5) =$
$200 + 100 = 300$

Filling Frames 2

1. $(10 + 7)(10) = 100 + 70 = 170$
2. $(10 + 3)(10 + 3) = 100 + 30 + 30 + 9 = 169$
3. $(10 + 8)(10 + 5) = 100 + 50 + 80 + 40 = 270$
4. $(10)(10 + 1) = 100 + 10 = 110$
5. $(10 + 4)(10 + 2) = 100 + 20 + 40 + 8 = 168$
6. $(10 + 9)(10 + 4) = 100 + 40 + 90 + 36 = 266$

FILLING FRAMES 1

Please cover the rectangles on the following pages with the fewest possible flats, sticks, and units. Begin at the origin with the flats. When no more flats fit, use sticks to build outward. When no more sticks fit, complete the frame with units.

Write a mathematical sentence for length x width = area using expanded notation and simplify the statement for the area.

Sketch a picture of each rectangle showing the four sections occupied by hundreds, tens, and ones. Write the value of each section next to that area. One problem has been done for you as an example.

1.

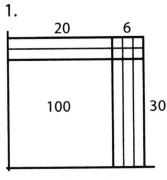

$(10 + 3)(10 + 2) = 100 + 20 + 30 + 6 = 156$

2.

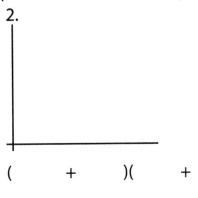

(+   )(+   ) =

      +   +   +   =

3.

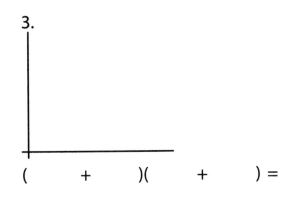

(+   )(+   ) =

      +   +   +   =

4.

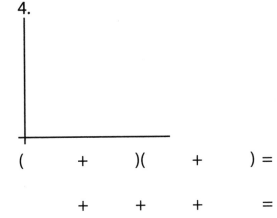

(+   )(+   ) =

      +   +   +   =

5.

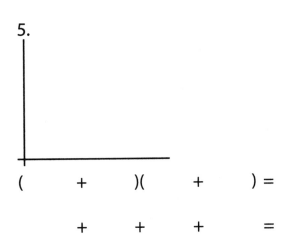

(+   )(+   ) =

      +   +   +   =

6.

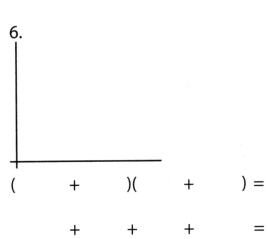

(+   )(+   ) =

      +   +   +   =

FILLING FRAMES 1

3

2

1

FILLING FRAMES 1

6

5

4

FILLING FRAMES²

Repeat the process from the first student sheets with the rectangles on the next two sheets.

1.

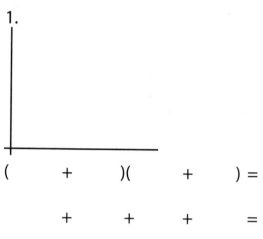

(+   )(+   ) =

    +   +   +   =

2.

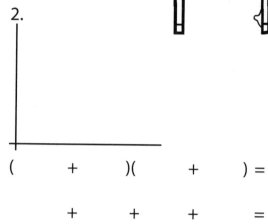

(+   )(+   ) =

    +   +   +   =

3.

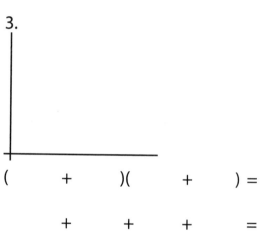

(+   )(+   ) =

    +   +   +   =

4.

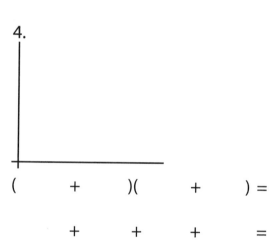

(+   )(+   ) =

    +   +   +   =

5.

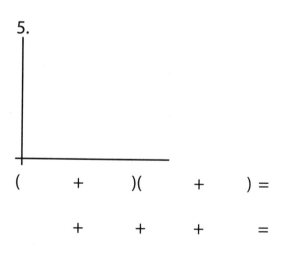

(+   )(+   ) =

    +   +   +   =

6.

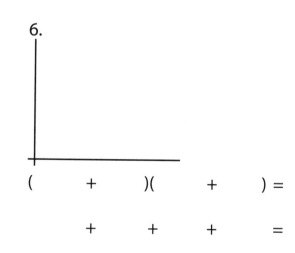

(+   )(+   ) =

    +   +   +   =

FILLING FRAMES²

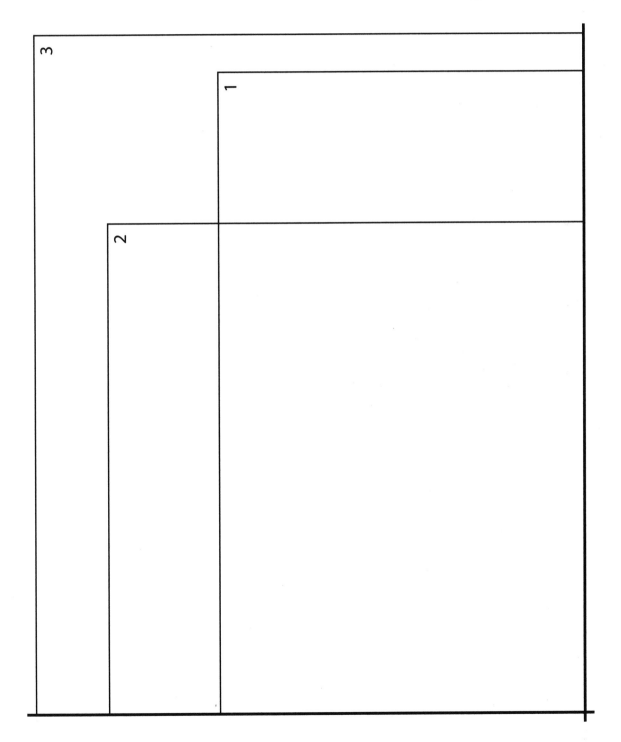

3

1

2

6

5

4

FILLING FRAMES

Connecting Learning

1. What did you discover about the way that the base-ten blocks are arranged in each rectangle?

2. How are these regions related to the expanded notation version of the multiplication problem?

3. How does this relationship help you to understand the meaning of the physical representation?

4. Was this method of looking at multiplication easier or harder for you than other methods you have done previously? Explain.

MODELS of Square Numbers

Topic
Distributive property and square numbers

Key Question
How can you compute the square of a number from a mental picture of its physical model?

Learning Goals
Students will:
- properly construct the models of square numbers,
- sketch representations of the models,
- accurately translate the models into the standard algebraic form of the distributive property,
- associate each term in the translated form with a component of the physical model and its representation, and
- see patterns of their own discovery in the sequence of consecutive square numbers to help them find the squares of the numbers.

Guiding Documents
Project 2061 Benchmark
- *Mathematical ideas can be represented concretely, graphically, and symbolically.*

Common Core State Standards for Mathematics*
- *Reason abstractly and quantitatively. (MP.2)*
- *Look for and make use of structure. (MP.7)*
- *Work with radicals and integer exponents. (8.EE.A)*

Math
Algebraic thinking
   distributive property
   expanded notation
Square numbers
Math patterns

Integrated Processes
Observing
Collecting and recording data

Materials
Student sheets
Base-ten blocks

Background Information
Language of Algebra

In this activity, the students interpret models of square numbers into the standard algebraic form of the distributive property. They begin at the manipulative level and move first to algebraic representation using squares, lines, and dots. From the model and representation they can determine the length, width, and area for expressing in standard algebraic form.

The purpose is to give meaning to each form and the interrelationships among them before introducing the process of multiplication at the abstract level. As students move to the abstract level, they should have a clear picture of what is transpiring.

This activity should help students develop a process whereby they can compute the square of a number from a mental picture of its model. For example, 17 x 17 will require 1 flat, 2 x 7 sticks, and 49 units that represent 100 + 140 + 49 = 289.

Management
1. Students need base-ten blocks in order to complete this activity. One set of blocks can be shared between a few students.
2. If students are not familiar with expanded notation, be sure to go over a few examples before they begin.

Procedure
1. Hand out the student sheets and base-ten blocks to each student and go over the instructions. You may wish to sketch a few numbers as a class so students are comfortable with the process.
2. Have students work individually or in small groups to fill the squares with blocks, sketch each figure, and determine the expanded notation for the area. Be sure that they describe the patterns that they discover in their models, sketches, and in the expanded notation.
3. When all students have completed both parts of the activity close with a time of class discussion and sharing.

Connecting Learning

1. What patterns do you see in the length and the width of the figures? [The number of flats stays the same, but the number of sticks increases by one in each successive square; the number of dots is always a square number; etc.]

2. What patterns do you see in the number sticks when the area is written in expanded notation? [Both stick values are the same, the values increase by 10 for each successive square, etc.]

3. What patterns do you see in the number of dots (ones) when the area is written in expanded notation? [The value is always a square number, it is the square of the number of lines in the length and width, etc.]

4. What would be the values for next two figures? (See *Solutions*.)

5. How did this activity help you understand the distributive property?

Solutions

The expanded notation formulas for each figure are given below including the answers for the next two squares.

Figure	Length	Width	Area
A	$(10 + 0)$	$(10 + 0)$	$= 100 + 0 + 0 + 0 = 100$
B	$(10 + 1)$	$(10 + 1)$	$= 100 + 10 + 10 + 1 = 121$
C	$(10 + 2)$	$(10 + 2)$	$= 100 + 20 + 20 + 4 = 144$
D	$(10 + 3)$	$(10 + 3)$	$= 100 + 30 + 30 + 9 = 169$
E	$(10 + 4)$	$(10 + 4)$	$= 100 + 40 + 40 + 16 = 196$
F	$(10 + 5)$	$(10 + 5)$	$= 100 + 50 + 50 + 25 = 225$
G	$(10 + 6)$	$(10 + 6)$	$= 100 + 60 + 60 + 36 = 256$
H	$(10 + 7)$	$(10 + 7)$	$= 100 + 70 + 70 + 49 = 289$
I	$(10 + 8)$	$(10 + 8)$	$= 100 + 80 + 80 + 64 = 324$
J	$(10 + 9)$	$(10 + 9)$	$= 100 + 90 + 90 + 81 = 361$

MODELS of Square Numbers

Fill each square with the fewest possible number of base ten blocks.

MODELS of Square Numbers

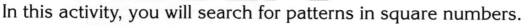

In this activity, you will search for patterns in square numbers.

a. Using the fewest possible number of base-ten blocks, fill each of the squares in turn.

b. Sketch a picture of the result.

c. Write an equation relating length, width, and area using expanded and exponential notation. Simplify the area expresssion.

A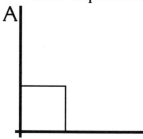

$(10 + 0) \ (10 + 0) = 100 + 0 + 0 + 0 = 100$
Length   Width   Area

B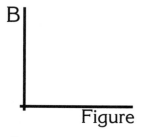

Figure

$(10 + 1) \ (10 + 1) = 100 + 10 + 10 + 1 = 121$
Length   Width   Area

C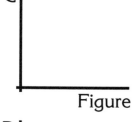

Figure

$(\quad + \quad)(\quad + \quad) = \quad + \quad + \quad + \quad = $
   Length   Width   Area

D

Figure

$(\quad + \quad)(\quad + \quad) = \quad + \quad + \quad + \quad = $
   Length   Width   Area

E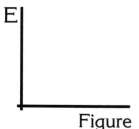

Figure

$(\quad + \quad)(\quad + \quad) = \quad + \quad + \quad + \quad = $
   Length   Width   Area

MODELS of Square Numbers

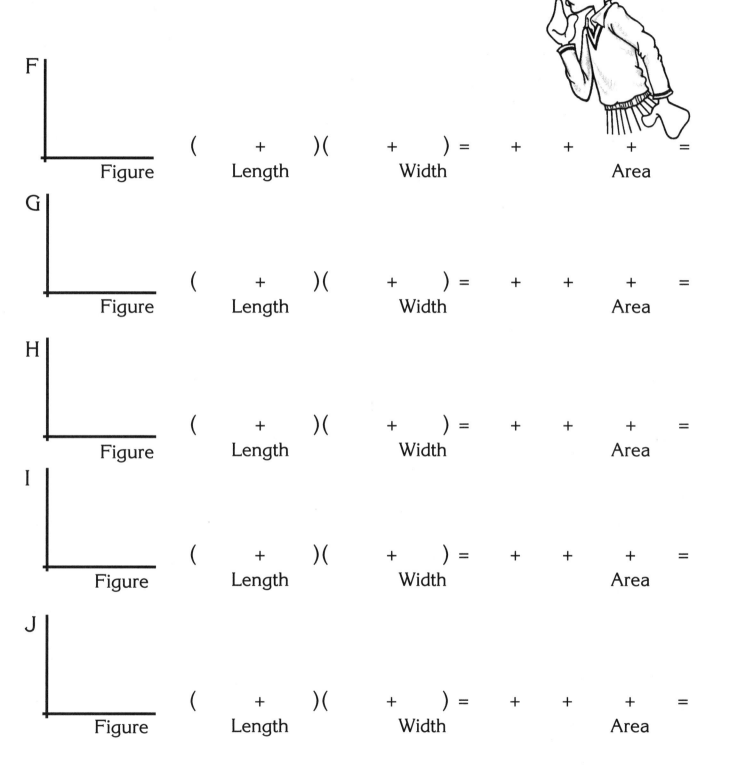

F
Figure
(　　　+　　　)(　　　+　　　) = 　　+　　+　　+　　 =
Length　　　　　Width　　　　　　　　　　Area

G
Figure
(　　　+　　　)(　　　+　　　) = 　　+　　+　　+　　 =
Length　　　　　Width　　　　　　　　　　Area

H
Figure
(　　　+　　　)(　　　+　　　) = 　　+　　+　　+　　 =
Length　　　　　Width　　　　　　　　　　Area

I
Figure
(　　　+　　　)(　　　+　　　) = 　　+　　+　　+　　 =
Length　　　　　Width　　　　　　　　　　Area

J
Figure
(　　　+　　　)(　　　+　　　) = 　　+　　+　　+　　 =
Length　　　　　Width　　　　　　　　　　Area

Please describe all the patterns you found in the completed models and in the pictures above on the back of this paper.

Connecting Learning

1. What patterns do you see in the length and the width of the figures?

2. What patterns do you see in the number sticks when the area is written in expanded notation?

3. What patterns do you see in the number of dots (ones) when the area is written in expanded notation?

4. What would be the values for next two figures?

5. How did this activity help you understand the distributive property?

CONSTRUCTIONS PLUS

Topic
Distributive property

Key Question
How can a set of base-ten blocks be arranged to form a rectangle?

Learning Goals
Students will:
- arrange sets of base-ten blocks to form rectangles, and
- write a mathematical sentence for length x width = area using expanded notation.

Guiding Documents
Project 2061 Benchmark
- *Mathematical ideas can be represented concretely, graphically, and symbolically.*

Common Core State Standards for Mathematics*
- *Model with mathematics. (MP.4)*
- *Use appropriate tools strategically. (MP.5)*
- *Look for and make use of structure. (MP.7)*
- *Use place value understanding and properties of operations to perform multi-digit arithmetic. (4.NBT.B)*

Math
Distributive property
Multiplication
Expanded notation
Base ten

Integrated Processes
Observing
Collecting and recording data
Generalizing

Materials
Base-ten blocks
Rulers, optional
Student sheets

Background Information
Certain sets of base-ten blocks, such as those given in this activity, can be used to build rectangles. *The rule that each flat must be as close to the origin as possible must always be followed.* If it is adhered to, the factors (length and width) will always be unique to that rectangle! Some students may have "tall" rectangles while others may have the same rectangles but theirs

are "wide." Examination will show that both are the same area and have the same dimensions, only the order of the factors is reversed. (You may want to take the time to let students test whether the factors for the rectangles are unique when the rule is followed. You may also wish to relate this to obtaining unique factors when algebraic equations are fully factored.)

When a set has one, two, or three flats, the only option is to arrange them in one row beginning at the origin. However, when the number of flats is four or any larger number, several options present themselves. Taking four as an example, the flats could form a one by four rectangle or a two by two rectangle. But a one by four rectangle violates the rule and is not permissible; the third and fourth flats in that array are not as close to the origin as possible.

It should be noted that the process in this activity is the inverse of that in *Filling Frames*. In *Filling Frames*, the rectangular frame determined which set of blocks would fill the frame; in this activity the blocks determine the frame that outlines the rectangle.

Management
1. Students will each need base-ten blocks to complete this activity. One set can be shared between two or three students.
2. Students should be familiar with how to use base-ten blocks and how to write an equation in expanded notation before beginning this activity. Be sure to emphasize the rule about building out from the origin so that each flat is as close to the origin as possible.
3. There are two parts to the student sheets: *Constructions Plus I* and *Constructions Plus II*. The second portion is for practice, and is not necessary if students have a good handle on the concept after the first half.
4. Students may want rulers to help them draw their sketches more neatly and accurately.

Procedure
1. Hand out the first student sheet and base-ten blocks to each student. Go over the instructions as a class and be sure students understand how to organize their flats, sticks, and units into the four groups. *It is possible to make at least one rectangle from each set of pieces listed. Create a rectangle on your desk using those pieces, then make a sketch of it in the space provided and record the area using expanded notation.*

2. If necessary, do an example together as a class, then have students work together in groups to complete the student sheet.
3. If desired, hand out the second student sheet for additional practice.
4. Once all students have completed the problems, close with a time of class discussion and sharing.

Connecting Learning

1. Was it difficult to determine how to arrange the pieces so that they formed a rectangle? Explain.
2. Is there more than one rectangle possible for any given set of base-ten blocks? [Not when the rules for arranging base-ten blocks are adhered to. While two rectangles are possible for several of the problems, they are merely rotations of each other, and have the same factors.]
3. How do you know which rectangle is correct? [In the cases where two orientations are possible, both are equally correct.]
4. If the rules for arranging base ten blocks were not followed, do you think there would be more than one solution for some problems? How would this be possible? (See *Solutions*.)
5. What did this activity teach you about multiplication?

Extension

Have students develop problems that have two, or even three, different solutions and exchange them with their classmates to solve.

Solutions

The solutions for both student pages are given. The first two problems have two solutions shown to give an example of the different orientations possible in many of the problems. While these solutions both have the same factors, the orientation makes them appear different when arranged with the base-ten blocks. Problem number six from the second student sheet is shown with two solutions, one that follows the rules for arranging base-ten blocks, and one that does not. The extension challenges students to create rectangles like this one, which have more than one set of factors.

Constructions Plus I

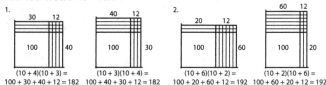

1. $(10 + 4)(10 + 3) =$
$100 + 30 + 40 + 12 = 182$
$(10 + 3)(10 + 4) =$
$100 + 40 + 30 + 12 = 182$

2. $(10 + 6)(10 + 2) =$
$100 + 20 + 60 + 12 = 192$
$(10 + 2)(10 + 6) =$
$100 + 60 + 20 + 12 = 192$

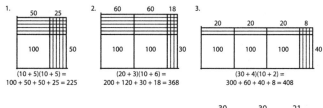

3. $(10 + 5)(10 + 1) =$
$100 + 10 + 50 + 5 = 165$

4. $(10 + 4)(10 + 4) =$
$100 + 40 + 40 + 16 = 196$

5. $(20 + 1)(10 + 3) =$
$200 + 60 + 10 + 3 = 273$

6. $(20 + 2)(20 + 5) =$
$400 + 100 + 40 + 10 = 550$

Constructions Plus II

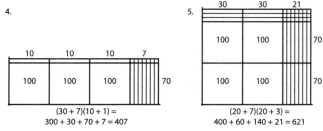

1. $(10 + 5)(10 + 5) =$
$100 + 50 + 50 + 25 = 225$

2. $(20 + 3)(10 + 6) =$
$200 + 120 + 30 + 18 = 368$

3. $(30 + 4)(10 + 2) =$
$300 + 60 + 40 + 8 = 408$

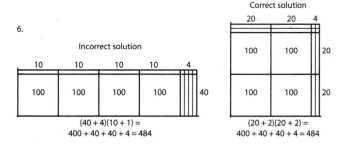

4. $(30 + 7)(10 + 1) =$
$300 + 30 + 70 + 7 = 407$

5. $(20 + 7)(20 + 3) =$
$400 + 60 + 140 + 21 = 621$

Correct solution

6.

Incorrect solution

$(40 + 4)(10 + 1) =$
$400 + 40 + 40 + 4 = 484$

$(20 + 2)(20 + 2) =$
$400 + 40 + 40 + 4 = 484$

CONSTRUCTIONS PLUS I

A specific set of base-ten blocks is listed for each problem. Your task is to form a rectangle using all of the blocks. Begin at the origin with the flats and build outward. The flats must be arranged in such a way that each is as close to the origin as possible. When you have used all of the flats, use the sticks. When those are all in place, finish by using units.

Sketch a picture of the resulting rectangle, showing the four sections occupied by hundreds, tens, and ones. Write the value of each section next to that area.

Write a mathematical sentence for length x width = area using expanded notation.

1. 1 flat, 7 sticks, 12 units

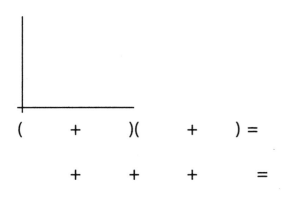

(____ + ____)(____ + ____) =

____ + ____ + ____ =

2. 1 flat, 8 sticks, 12 units

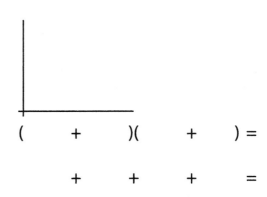

(____ + ____)(____ + ____) =

____ + ____ + ____ =

3. 1 flat, 6 sticks, 5 units

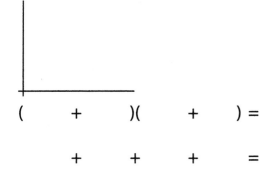

(____ + ____)(____ + ____) =

____ + ____ + ____ =

4. 1 flat, 8 sticks, 16 units

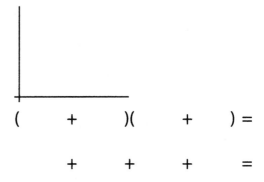

(____ + ____)(____ + ____) =

____ + ____ + ____ =

5. 2 flats, 7 sticks, 3 units

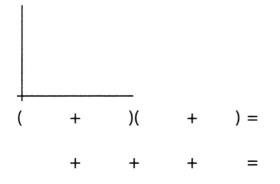

(____ + ____)(____ + ____) =

____ + ____ + ____ =

6. 4 flats, 14 sticks, 10 units

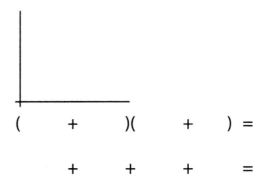

(____ + ____)(____ + ____) =

____ + ____ + ____ =

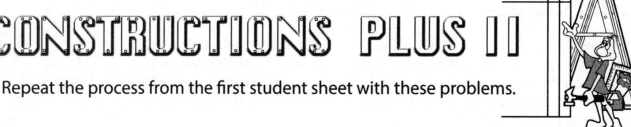

CONSTRUCTIONS PLUS II

Repeat the process from the first student sheet with these problems.

1. 1 flat, 10 sticks, 25 units

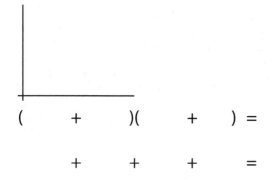

(+   )(+   ) =

    +   +   +   =

2. 2 flats, 15 sticks, 18 units

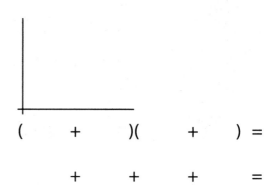

(+   )(+   ) =

    +   +   +   =

3. 3 flats, 10 sticks, 8 units

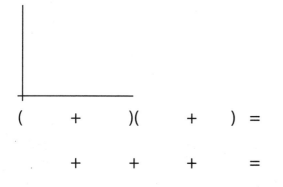

(+   )(+   ) =

    +   +   +   =

4. 3 flats, 10 sticks, 7 units

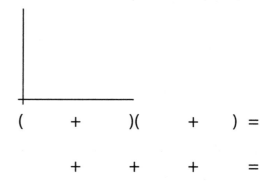

(+   )(+   ) =

    +   +   +   =

5. 4 flats, 20 sticks, 21 units

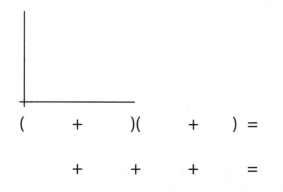

(+   )(+   ) =

    +   +   +   =

6. 4 flats, 8 sticks, 4 units

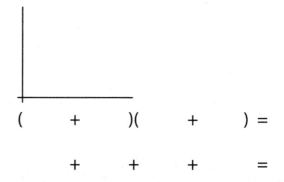

(+   )(+   ) =

    +   +   +   =

CONSTRUCTIONS PLUS

Connecting Learning

1. Was it difficult to determine how to arrange the pieces so that they formed a rectangle? Explain.

2. Is there more than one rectangle possible for any given set of base-ten blocks?

3. How do you know which rectangle is correct?

4. If the rules for arranging base ten blocks were not followed, do you think there would be more than one solution for some problems? How would this be possible?

5. What did this activity teach you about multiplication?

Decimal Numbers and the Distributive Property

S tudents will acquire a deeper understanding of the meaning and relative magnitude of decimal fractions as they progress through the activities in *Part Four.* A primary objective is for students to build *mental images* of the interrelationships among ones, tenths, and hundredths by manipulating base-ten blocks and picturing the results.

Students will learn to multiply decimal fractions using the area model, express the multiplication of decimal fractions by dimension and area product, recognize the similarities between whole numbers and decimal fractions, and use the distributive property. They will develop place value insights through the application of the distributive property in the multiplication of decimal fractions at three levels: concrete, representational, and abstract. Each level makes its specific contribution to greater understanding.

It is important to recognize the contribution that is made by using both "doing" and "undoing" processes. These inverse processes combine to build meaning and deepen understanding. At times, concrete models or their representations are translated into the abstract. At other times, the inverse occurs as abstract expressions and equations are translated into concrete models or pictures.

The experiences in this section open the door for a variety of teacher-directed activities involving the representation of decimals. Students can be asked to name the decimal fraction that corresponds to a set of blocks displayed on the overhead. Alternatively, each student can be given a set of blocks and asked to display models of such numbers as 2.43, 1.05, etc.

Teacher-directed activities can also be used to extend activities involving the multiplication of decimal fractions. Students can be asked to model a product and to determine the dimensions or factors involved. An alternative is to display a picture of a product and ask students to determine the dimensions and product and enter them into the appropriate quadratic equation.

The advantage of such teacher-directed activities is that they can be inserted anywhere into the day's schedule for whatever length of time is available. This provides periodic reinforcement and review.

From Tens to Tenths
This activity connects the place value concepts underlying whole number multiplication to an understanding of decimal fraction place value. Students look at pictorial representations of decimal fractions and interpret them into abstract expressions.

More From Tens to Tenths
Using the inverse process of that in *From Tens to Tenths,* students construct and sketch representations of decimal fractions.

From Tens to Tenths—Again
The multiplication of decimal fractions is introduced. It builds on the understandings developed in work with whole numbers. Students are given the dimensions of a rectangle in abstract form and are asked to build a model and draw a sketch of the result. They are asked to identify the components in the model in terms of ones, tenths, and hundredths, and write an expression representing the sum of the components.

More From Tens to Tenths—Again
Students write the multiplication problem illustrated by each rectangle, identifying and relating the length, width, and area.

Filling Frames—From Tens to Tenths
Students fill rectangles using the fewest number of blocks. They employ the same procedure they used with whole numbers by building with blocks representing ones first and constructing outward using blocks modeling tenths and hundredths. The model is then translated into a quadratic equation relating length, width, and area.

FROM Tens to Tenths

Topic
Multiplying decimals

Key Questions
How can you use base-ten materials to show multiplication with decimals?

Learning Goals
Students will:
- learn to multiply decimals using the area model,
- express the multiplication of decimals by dimension and area product,
- recognize the similarities between whole numbers and decimal numbers, and
- use the distributive property.

Guiding Documents
Project 2061 Benchmarks
- *Add, subtract, multiply, and divide whole numbers mentally, on paper, and with a calculator.*
- *Express numbers like 100, 1000, and 1,000,000 as powers of ten*

Common Core State Standards for Mathematics*
- *Model with mathematics. (MP.4)*
- *Use appropriate tools strategically. (MP.5)*
- *Look for and make use of structure. (MP.7)*
- *Perform operations with multi-digit whole numbers and with decimals to hundredths. (5.NBT.B)*

Math
Multiplication
  decimals
  sums of partial products using an area model
Place value
Distributive property

Integrated Processes
Observing
Comparing and contrasting
Generalizing

Materials
Base-ten blocks
Blank rectangles from *Filling Frames*
Student sheets

Background Information
By the time students begin this activity, they should have experienced multiplying with base-ten materials using two-digit whole numbers. They should have a basic proficiency using base-ten blocks—both naming numbers, and using the area model to multiply. This activity will connect the place value concepts underlying whole number multiplication to an understanding of decimal fraction place value.

Using base-ten blocks, students can multiply 23 x 12 in this manner:

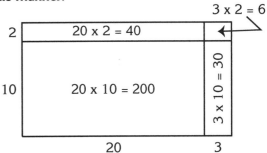

They have recognized that, using the dimensions of 23 (or 20 + 3) by 12 (or 10 + 2), the resulting rectangle is divided into four regions. (See *Multiplying with Tens, Building Rectangles*, and *Picturing Rectangles*.) The students have learned how to construct a rectangle and name its area given its dimensions. After constructing the rectangles, they have labeled them using the dimensions—(20 + 3) and (10 + 2)—and the partial products—200, 40, 30, 6—of the four internal areas.

Many math books require students to express a relationship of sums of products and products of sums. This is a verbal way to describe the distributive property. The dimensions of the rectangle are expressed as sums (20 + 3) and (10 + 2). To find the area of the rectangle, we multiply the two sides, thus a product of sums (20 + 3)(10 + 2). The area can be expressed as the sum of the four partial products: (20 x 10) + (20 x 2) + (3 x 10) + (3 x 2). Therefore, we calculate the sum of products 200 + 40 + 30 + 6 to find the total area. Students who have calculated and manipulated symbols and numbers to practice the distributive property realize that:

$$23 \times 12 = (20 + 3)(10 + 2)$$
$$= (20 \times 10) + (20 \times 2) + (3 \times 10) + (3 \times 2)$$
$$= 200 + 40 + 30 + 6$$
$$= 276$$

We will relate this model to decimals, since picturing a decimal is confusing for many students. We will construct models of decimals and multiply them to gain experience and facility in multiplying decimals.

We will rename the base-ten blocks as follows:

The flat (10 x 10 = 100)	will be 1 (or 1 x 1);
the stick (1 x 10 = 10)	will be 0.1 (or 1 x 0.1);
and the unit cubes (1 x 1 = 1)	will be 0.01 (or 0.1 x 0.1).

Using this system, the number 1.24 can be represented as follows:

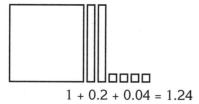

$$1 + 0.2 + 0.04 = 1.24$$

This ability to rethink names of base-ten blocks aids in bridging the gap from arithmetic to algebra. We ask students to take a leap of faith when we begin to use variables. Their ability to rename base-ten blocks will link to the naming of algebra blocks (when we call a length *n*).

Management

1. Students should work together in small groups on this activity. Each group will need a set of base-ten blocks.
2. This activity contains multiple experiences using the base-ten blocks in a decimal context. Depending on the level of your students, you may wish to spread this activity over several days or only use the parts of it best suited to your class.
3. Allow a few minutes for students to build with blocks before you begin the activity. This prevents further block building when you are trying to teach.
4. Review base-ten multiplication before beginning the new concept of decimal multiplication.
5. At first, require the students to build each figure before calculating the area. Students should get into the habit of always building, sketching, and labeling the rectangles. After gaining experience, some students will be able to do the multiplication without the manipulatives; however, their understanding will deepen if they first use the blocks.

Procedure

1. Review decimal place value briefly so that students are refreshed on tens, tenths, and hundredths.
2. Have students get into their groups and distribute the base-ten blocks. Allow time for students to play with the blocks before continuing.
3. Introduce the values of the base-ten blocks (flat = 1, stick = 0.1, and unit cube = 0.01).
4. Hand out the first student sheet, *From Tens to Tenths*, and have students work to translate the pictures into numeric form.
5. When groups are finished, hand out *More From Tens to Tenths*. Be sure that students build each rectangle before sketching it.
6. When students are ready to move on, hand out the sheets *From Tens to Tenths—Again* and *More From Tens to Tenths—Again*. These sheets will move students into decimal multiplication using the area model.

7. Be sure that students understand the process of sketching and labeling the rectangles, and then writing the partial products and the total area.
8. As they work on these pages, circulate among the groups, checking for understanding.
9. To reinforce the concept of multiplication using the area model, hand out the blank rectangles from *Filling Frames*. Have students create each of the rectangles on that sheet using their base-ten blocks. The final student sheet, *Filling Frames—From Tens to Tenths* can be used to record the data in decimal form.

Connecting Learning

1. What is place value? What are the similarities between tens and tenths?
2. Which is greater 0.1 or 0.01? [0.1] Why? [0.1 is one tenth, while 0.01 is one hundredth]
3. How do you show multiplication with decimals by sketching? [You draw each of the four partial products. (See *Solutions* for examples.)]
4. How is multiplying with decimals like multiplying with whole numbers?

Extensions

1. Repeat this activity using negative powers of ten to develop an understanding of negative exponents. Hence $0.1 = 10^{-1}$, $0.01 = 10^{-2}$, and $0.001 = 10^{-3}$.
2. Challenge students to write the number values using expanded notation.
3. Compare the numeric solutions of 14 x 13 and 1.4 x 1.3. What are the similarities and differences?
4. When we teach the skill of decimal multiplication, we ask students to multiply as usual, then count the digits in the factors which are on the right of the decimal. Then we count from the "back" of a number to put the decimal point in the product. How is this concept illustrated in the area model of multiplying with decimals? Where do we get the notion of moving a decimal point?

Solutions

The solutions from each of the student sheets are given below.

From Tens to Tenths

1. 1.25
2. 1.44
3. 2.15
4. 3.51
5. 2.64
6. 3.73
7. 2.03
8. 2.87

More From Tens to Tenths

1.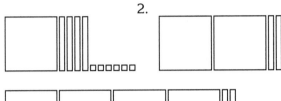
2.

3.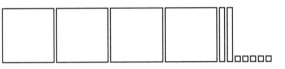

4.
5.

6.

7.

8.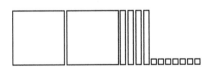

9.

10.

From Tens to Tenths—Again

2. $1.3 \times 1.5 = 1 + .5 + .3 + .15 = 1.95$

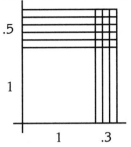

3. $1.6 \times 1.2 = 1 + .2 + .6 + .12 = 1.92$

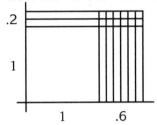

4. $1.4 \times 2.3 = 2 + .3 + .8 + .12 = 3.22$

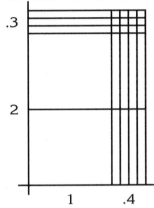

5. $2.3 \times 4.1 = 8 + .2 + 1.2 + .03 = 9.43$

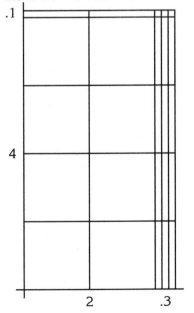

6. 4.2 x 2.1 = 8 + .4 + .4 + .02 = 8.82

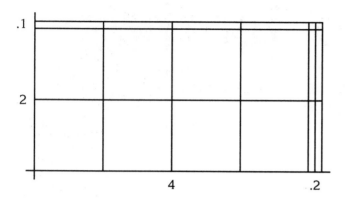

.1

2

4   .2

More From Tens to Tenths—Again
1. 1.1 x 1.3 =
   1 + .3 + .1 + .03 = 1.43
2. 2.4 x 1.2 =
   2 + .4 + .4 + .08 = 2.88
3. 2.2 x 2.1 =
   4 + .2 + .4 + .02 = 4.62
4. 3.1 x 1.3 =
   3 + .9 + .1 + .03 = 4.03
5. 1.3 x 2.2 =
   2 + .2 + .6 + .06 = 2.86
6. 2.5 x 4.1 =
   8 + .2 + 2.0 + .05 = 10.25

Filling Frames—From Tens to Tenths

	Sketch	Dimensions	Partial Products	Area
1		1.3 x 1.2	1 + .2 + .3 + .06	1.56
2		1.7 x 1.3	1 + .3 + .7 + .21	2.21
3		1.9 x 1.5	1 + .5 + .9 + .45	2.85
4		1.6 x 1.0	1 + .6	1.6
5		2.1 x 1.1	2 + .2 + .1 + .01	2.31
6		2.0 x 1.5	2 + 1.0	3

FROM Tens to Tenths

Give the number value for each of the pictures below, assuming that the base-ten blocks have the following values:

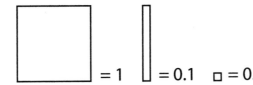

$\square = 1$ $\mid = 0.1$ $\square = 0.01$

1.

5.

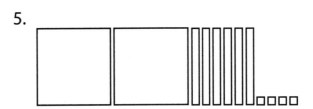

2.

6.

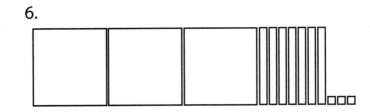

3.

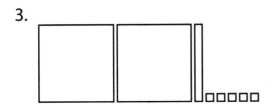

7.

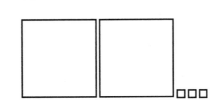

8.

4.

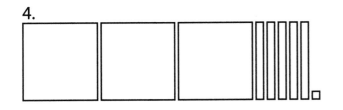

MULTIPLICATION THE ALGEBRA WAY   103   © 2012 AIMS Education Foundation

MORE from Tens to Tenths

Build and sketch each of the numbers below, assuming these values for the base-ten blocks:

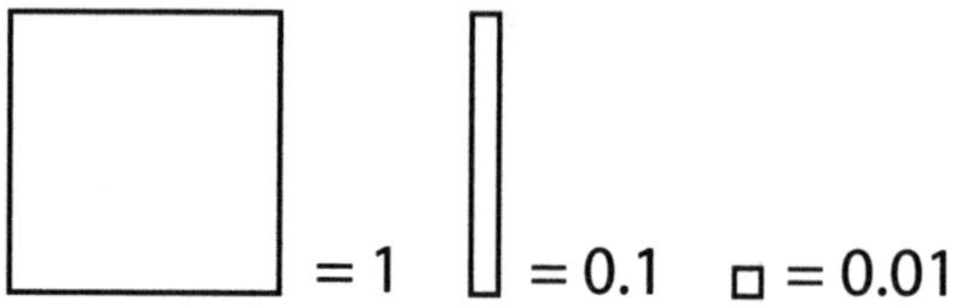

$\square$ = 1 $\vert$ = 0.1 $\square$ = 0.01

1. 1.46

2. 2.31

3. 4.25

4. 0.23

5. 3.71

6. 0.72

7. 3.06

8. 2.47

9. 1.52

10. 0.64

Tens to Tenths *Again*

Show these multiplication problems by building them, sketching them on the grid, and labeling the dimensions. Show the four partial products and the total area. The first problem has been done for you as an example.

Assume the following values for the base-ten blocks:

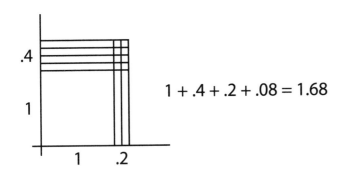

= 1   = 0.1   □ = 0.01

1. $1.2 \times 1.4 =$

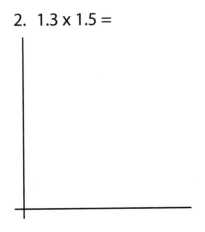

.4

1

$1 + .4 + .2 + .08 = 1.68$

1   .2

2. $1.3 \times 1.5 =$

3. $1.6 \times 1.2 =$

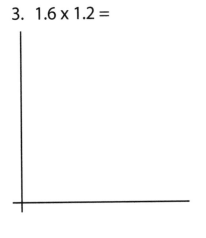

4. $1.4 \times 2.3 =$

5. $2.3 \times 4.1 =$

6. $4.2 \times 2.1 =$

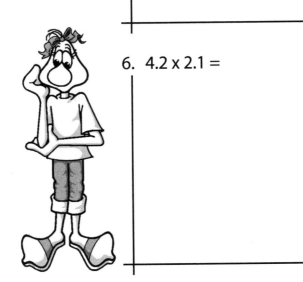

Write the multiplication problem that is illustrated by each rectangle below. Label the dimensions on the grid and show the four partial products and the total area.

Assume the following values for the base-ten blocks:

= 1   = 0.1   □ = 0.01

1.

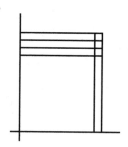

2.

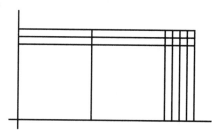

3.

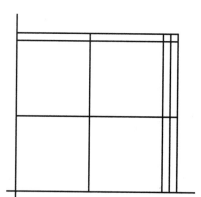

4.

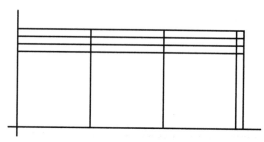

5.

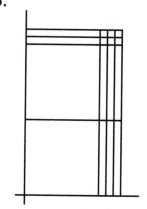

6.

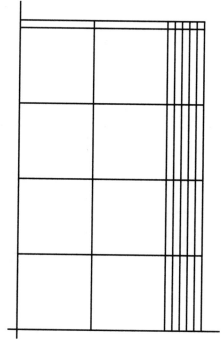

Filling Frames From Tens to Tenths

Use the *Filling Frames 1* sheets and build each of the rectangles with your base-ten blocks. As you build a rectangle, sketch it in the table and fill in the dimensions, partial products, and the area.

Assume the following values for the base-ten blocks:

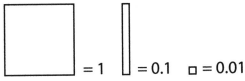 = 1   = 0.1   □ = 0.01

	Sketch	Dimensions	Partial Products	Area
1				
2				
3				
4				
5				
6				

Connecting Learning

1. What is place value? What are the similarities between tens and tenths?

2. Which is greater 0.1 or 0.01? Why?

3. How do you show multiplication with decimals by sketching?

4. How is multiplying with decimals like multiplying with whole numbers?

Multiplication of Fractions

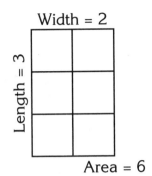

Width = 2

Length = 3

Area = 6

I n mathematics, multiplication is generally represented by two mutually perpendicular measures whose product is a measure of area. For example, the length and width of a rectangle are mutually perpendicular and their product is a measure of the area of the rectangle.

In the unit square model, this same perpendicular relationship exists. To represent the denominator of the first factor, lines are drawn parallel to one side to create the number of congruent subdivisions named by its denominator. The numerator is represented by shading in the number of these spaces it names.

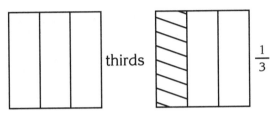

thirds $\frac{1}{3}$

To multiply by the second factor, the same procedure is followed except that the dividing lines are drawn perpendicular to those in the first instance.

The total number of subdivisions formed represents the denominator of the product. The number of subdivisions in which the numerators overlap names the numerator of the product.

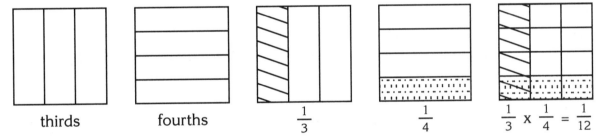

thirds   fourths $\frac{1}{3}$ $\frac{1}{4}$ $\frac{1}{3}$ x $\frac{1}{4}$ = $\frac{1}{12}$

Included is a set of Fraction Squares that help model the multiplication of fractions. This set should be run on transparencies and then cut out. The set consists of transparent unit squares that show the fractions $\frac{1}{2}$, $\frac{1}{3}$, $\frac{2}{3}$, $\frac{1}{4}$, $\frac{3}{4}$, $\frac{2}{5}$, $\frac{3}{5}$, $\frac{1}{3}$, and $\frac{5}{6}$. The multiplication of two fractions is modeled by superimposing the model of one fraction on top of the model of a second fraction with dividing lines running perpendicular to the first. This set is excellent for overhead use in conjunction with any activity in this part.

To meet the specific needs of students, teachers may wish to reorder the sequence in which these activities are used.

In *Fraction Factors,* students are asked to draw pictures of the two fraction factors. Reference to the AIMS Fraction Squares is particularly effective at this stage. The product is read from the completed picture and recorded to complete the sentence.

Product Pictures require reading the pictures to determine the factors and the products. *Fraction Frames* has students identify the factors and product, find, and complete the matching sentence. This activity works well as an assessment.

The use of the distributive property in the multiplication of mixed numbers is addressed in *Expanding on Fractions.*

Fraction Squares

Copy the two pages of Unit Squares onto transperency film. Cut out squares. To model the multiplication of fractions, overlay the appropriate dots and stripe squares.

110

Fraction Squares

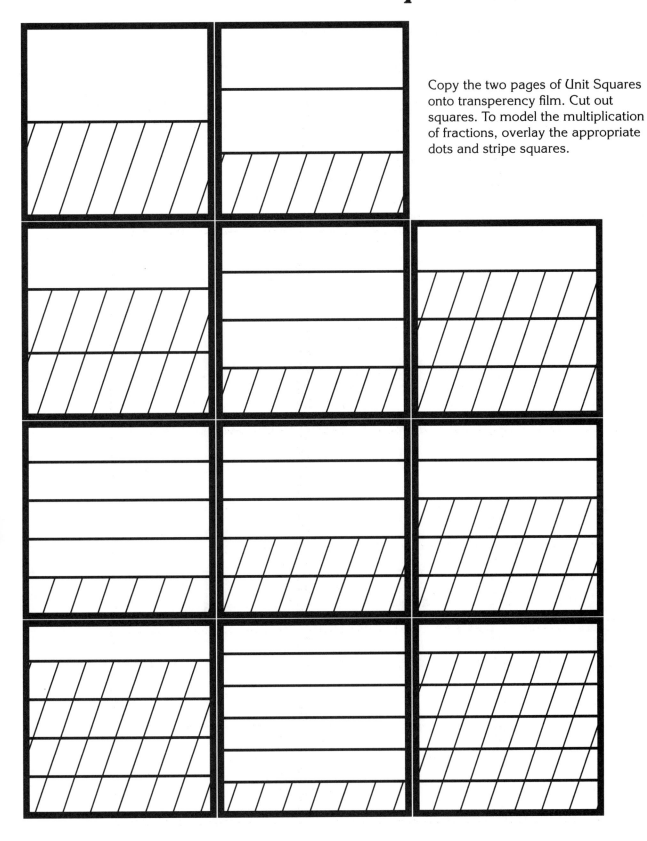

Copy the two pages of Unit Squares onto transperency film. Cut out squares. To model the multiplication of fractions, overlay the appropriate dots and stripe squares.

Fraction Factors

a. Please represent the problem graphically.
b. Determine the product from the drawing and complete the sentence.

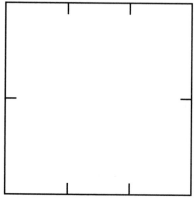 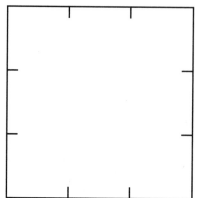 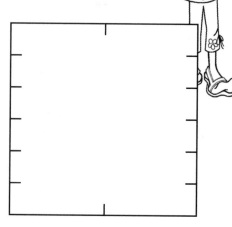

$\frac{1}{3}$ X $\frac{1}{2}$ = $\frac{1}{3}$ X $\frac{2}{3}$ = $\frac{1}{2}$ X $\frac{1}{6}$ =

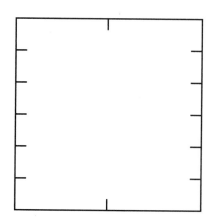 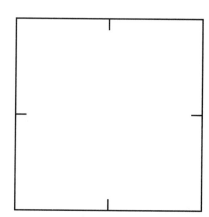

$\frac{1}{2}$ X $\frac{2}{3}$ = $\frac{1}{2}$ X $\frac{5}{6}$ = $\frac{1}{2}$ X $\frac{1}{2}$ =

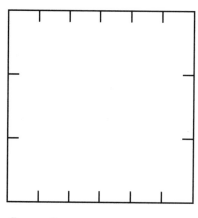 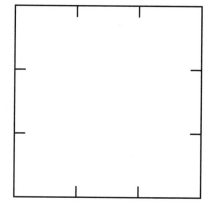

$\frac{1}{3}$ X $\frac{1}{3}$ = $\frac{5}{6}$ X $\frac{2}{3}$ = $\frac{2}{3}$ X $\frac{2}{3}$ =

Fraction Factors

a. Please find the following products.
b. Check your result using the AIMS Fraction Squares.

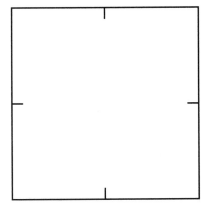

$\frac{1}{2}$ X $\frac{1}{2}$ =

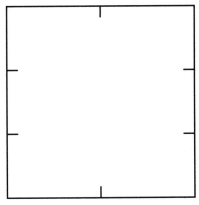

$\frac{1}{2}$ X $\frac{2}{3}$ =

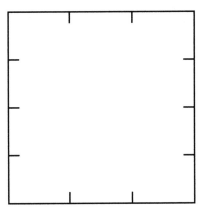

$\frac{1}{3}$ X $\frac{3}{4}$ =

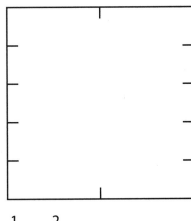

$\frac{1}{2}$ X $\frac{2}{5}$ =

$\frac{2}{3}$ X $\frac{1}{4}$ =

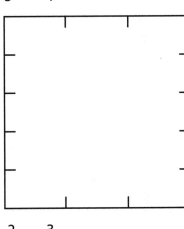

$\frac{2}{3}$ X $\frac{3}{5}$ =

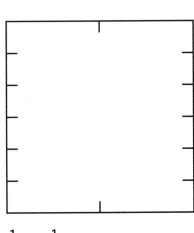

$\frac{1}{2}$ X $\frac{1}{6}$ =

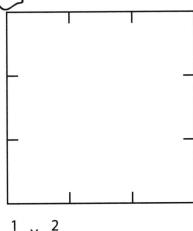

$\frac{1}{3}$ X $\frac{2}{3}$ =

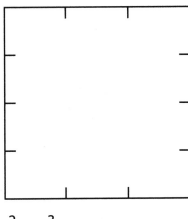

$\frac{2}{3}$ X $\frac{3}{4}$ =

Product Pictures

Determine the factors and products in the pictures.
Write the number sentence that describes each.

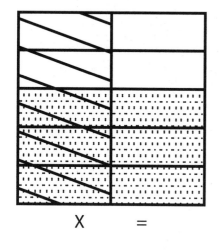

X ___ = ___

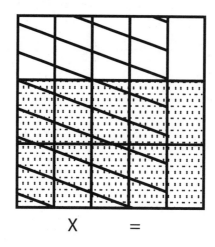

X ___ = ___

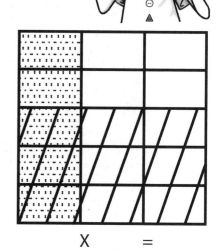

X ___ = ___

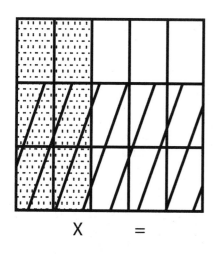

X ___ = ___

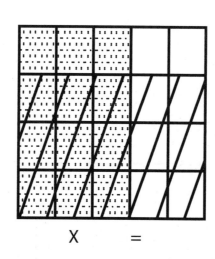

X ___ = ___

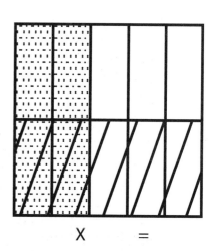

X ___ = ___

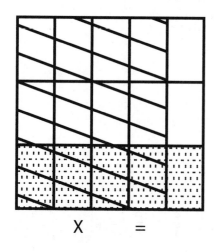

X ___ = ___

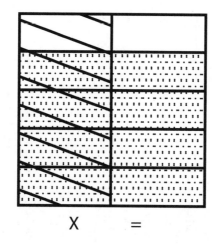

X ___ = ___

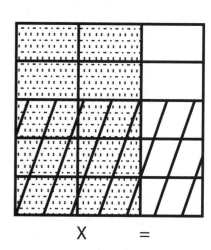

X ___ = ___

Fraction Frames

Please match each picture to a sentence below. Then complete the sentence by writing the answer.

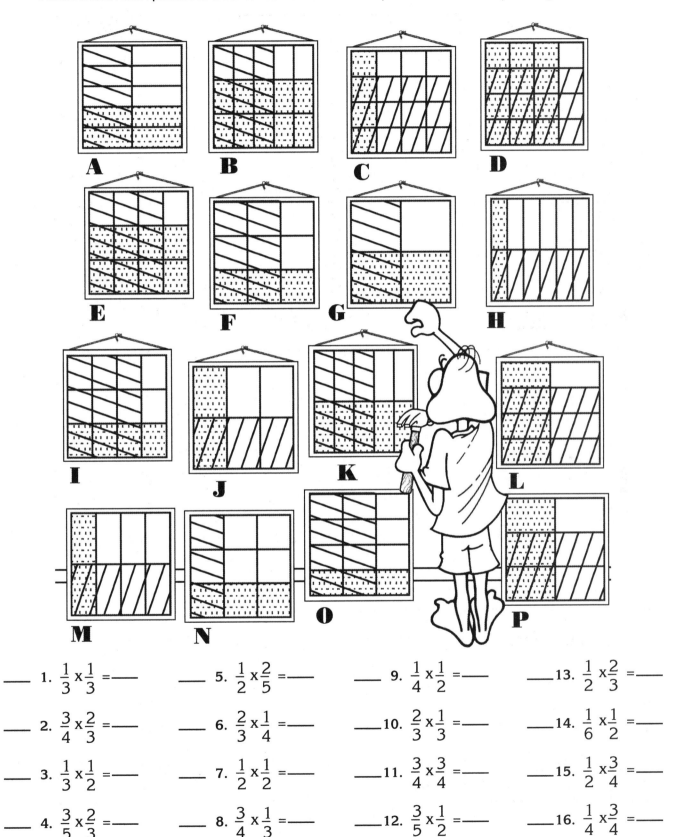

___ 1. $\frac{1}{3} \times \frac{1}{3} =$ ___

___ 2. $\frac{3}{4} \times \frac{2}{3} =$ ___

___ 3. $\frac{1}{3} \times \frac{1}{2} =$ ___

___ 4. $\frac{3}{5} \times \frac{2}{3} =$ ___

___ 5. $\frac{1}{2} \times \frac{2}{5} =$ ___

___ 6. $\frac{2}{3} \times \frac{1}{4} =$ ___

___ 7. $\frac{1}{2} \times \frac{1}{2} =$ ___

___ 8. $\frac{3}{4} \times \frac{1}{3} =$ ___

___ 9. $\frac{1}{4} \times \frac{1}{2} =$ ___

___ 10. $\frac{2}{3} \times \frac{1}{3} =$ ___

___ 11. $\frac{3}{4} \times \frac{3}{4} =$ ___

___ 12. $\frac{3}{5} \times \frac{1}{2} =$ ___

___ 13. $\frac{1}{2} \times \frac{2}{3} =$ ___

___ 14. $\frac{1}{6} \times \frac{1}{2} =$ ___

___ 15. $\frac{1}{2} \times \frac{3}{4} =$ ___

___ 16. $\frac{1}{4} \times \frac{3}{4} =$ ___

Solutions

Fraction Factors

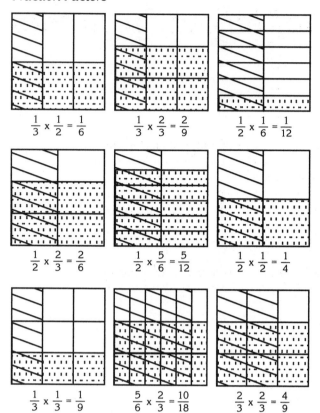

$$\frac{1}{3} \times \frac{1}{2} = \frac{1}{6}$$ $$\frac{1}{3} \times \frac{2}{3} = \frac{2}{9}$$ $$\frac{1}{2} \times \frac{1}{6} = \frac{1}{12}$$

$$\frac{1}{2} \times \frac{2}{3} = \frac{2}{6}$$ $$\frac{1}{2} \times \frac{5}{6} = \frac{5}{12}$$ $$\frac{1}{2} \times \frac{1}{2} = \frac{1}{4}$$

$$\frac{1}{3} \times \frac{1}{3} = \frac{1}{9}$$ $$\frac{5}{6} \times \frac{2}{3} = \frac{10}{18}$$ $$\frac{2}{3} \times \frac{2}{3} = \frac{4}{9}$$

Product Pictures

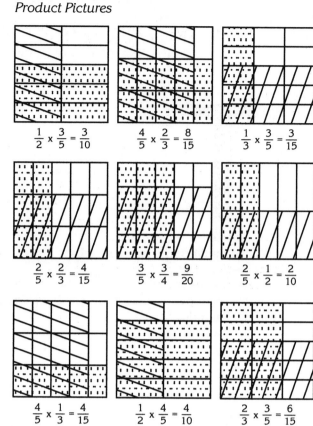

$$\frac{1}{2} \times \frac{3}{5} = \frac{3}{10}$$ $$\frac{4}{5} \times \frac{2}{3} = \frac{8}{15}$$ $$\frac{1}{3} \times \frac{3}{5} = \frac{3}{15}$$

$$\frac{2}{5} \times \frac{2}{3} = \frac{4}{15}$$ $$\frac{3}{5} \times \frac{3}{4} = \frac{9}{20}$$ $$\frac{2}{5} \times \frac{1}{2} = \frac{2}{10}$$

$$\frac{4}{5} \times \frac{1}{3} = \frac{4}{15}$$ $$\frac{1}{2} \times \frac{4}{5} = \frac{4}{10}$$ $$\frac{2}{3} \times \frac{3}{5} = \frac{6}{15}$$

Fraction Factors

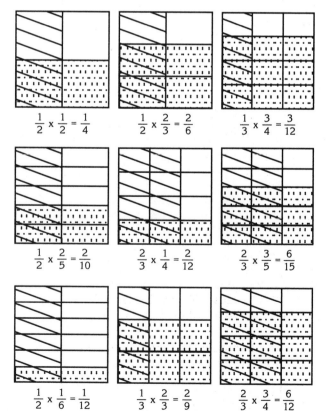

$$\frac{1}{2} \times \frac{1}{2} = \frac{1}{4}$$ $$\frac{1}{2} \times \frac{2}{3} = \frac{2}{6}$$ $$\frac{1}{3} \times \frac{3}{4} = \frac{3}{12}$$

$$\frac{1}{2} \times \frac{2}{5} = \frac{2}{10}$$ $$\frac{2}{3} \times \frac{1}{4} = \frac{2}{12}$$ $$\frac{2}{3} \times \frac{3}{5} = \frac{6}{15}$$

$$\frac{1}{2} \times \frac{1}{6} = \frac{1}{12}$$ $$\frac{1}{3} \times \frac{2}{3} = \frac{2}{9}$$ $$\frac{2}{3} \times \frac{3}{4} = \frac{6}{12}$$

Fraction Frames

__N__ 1. $\frac{1}{3} \times \frac{1}{3} = \frac{1}{9}$ __A__ 5. $\frac{1}{2} \times \frac{2}{5} = \frac{2}{10}$ __M__ 9. $\frac{1}{4} \times \frac{1}{2} = \frac{1}{8}$ __P__ 13. $\frac{1}{2} \times \frac{2}{3} = \frac{2}{6}$

__E__ 2. $\frac{3}{4} \times \frac{2}{3} = \frac{6}{12}$ __O__ 6. $\frac{2}{3} \times \frac{1}{4} = \frac{2}{12}$ __F__ 10. $\frac{2}{3} \times \frac{1}{3} = \frac{2}{9}$ __H__ 14. $\frac{1}{6} \times \frac{1}{2} = \frac{1}{12}$

__J__ 3. $\frac{1}{3} \times \frac{1}{2} = \frac{1}{6}$ __G__ 7. $\frac{1}{2} \times \frac{1}{2} = \frac{1}{4}$ __D__ 11. $\frac{3}{4} \times \frac{3}{4} = \frac{9}{16}$ __L__ 15. $\frac{1}{2} \times \frac{3}{4} = \frac{3}{8}$

__B__ 4. $\frac{3}{5} \times \frac{2}{3} = \frac{6}{15}$ __I__ 8. $\frac{3}{4} \times \frac{1}{3} = \frac{3}{12}$ __K__ 12. $\frac{3}{5} \times \frac{1}{2} = \frac{3}{10}$ __C__ 16. $\frac{1}{4} \times \frac{3}{4} = \frac{3}{16}$

The Distributive Property and the Multiplication of Mixed Numbers

The distributive property of multiplication over addition underlies all multiplication operations in arithmetic and algebra. Yet, the standard multiplication algorithm used in arithmetic masks the distributive property, hides the "ten-ness" of our numeration system, and is not transferable into algebra. It appears to be an efficient "bag of tricks" that somehow leads to the right answer.

The standard algorithm for the multiplication of mixed numbers suffers from similar deficiencies. It there a better alternative?

The answer is yes! It lies in using the algebraic approach to the distributive property of multiplication over addition. Does this alternative apply to mixed numbers as well as whole numbers and decimals? Again, the answer is yes.

Algebraically, the product of x + y and 2x is found as shown below:

$$2x\,(x + y) = 2x^2 + 2xy$$

The product of two binomials is found as follows:

$$(x + y)(2x + 3y) = 2x^2 + 3xy + 2xy + 3y^2$$

With whole numbers, this gives rise to two algorithms—the *algebraic form* and *display multiplication*. Consider the problem 47 x 89.

Algebraic form: $(40 + 7)(80 + 9) = 3200 + 360 + 560 + 63 = 4183$

Display Method:

```
        47
      x 89
        63   (9 x 7)
       360   (9 x 40)
       560   (80 x 7)
      3200   (80 x 40)
      4183
```

Both make an important contribution to understanding in that they bring out the "ten-ness" of our numeration system. In the algebraic form, units, tens, hundreds, etc., are separated. The display method requires constant attention to place value, thus emphasizing the ten-ness of the numeration system.

It is interesting to note that both methods yield the same sub-products albeit in a different order. Each is straightforward, devoid of shortcuts that involve meaningless procedures.

A companion element consistent with these algorithms is found in the graphical representation of multiplication shown below. These pictures are powerful for clarifying concepts and processes and reinforcing meaning. Modeling the multiplication of multi-digit numbers and fractions consists of

1. representing addends within factors as line segments laid end to end;
2. representing multiplication as perpendicular array of the factors, specifically as the dimensions of a rectangle; and
3. completing the rectangle, whose area represents the product.

This illustration is similar to that found in many algebra texts. It pictures $(x + y)(y + x)$.

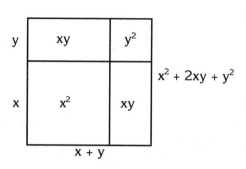

$$x^2 + 2xy + y^2$$

The parallel in arithmetic for 15 x 15 is shown at the right.

The application of the algebraic algorithm to the multiplication of mixed numbers together with its graphical representation is shown in the next examples. By using a grid in which the area of a 12 x 12 region represents 1, the pictures of mixed fractions with denominators of 2, 3, 4, 6, and 12 (factors of 12) are readily displayed.

Note that the areas of the four regions correspond to the four sub-products created when the algebraic algorithm is used.

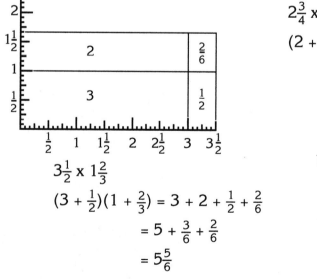

$3\frac{1}{2}$ x $1\frac{2}{3}$

$(3 + \frac{1}{2})(1 + \frac{2}{3}) = 3 + 2 + \frac{1}{2} + \frac{2}{6}$

$= 5 + \frac{3}{6} + \frac{2}{6}$

$= 5\frac{5}{6}$

$2\frac{3}{4}$ x $1\frac{1}{3}$

$(2 + \frac{3}{4})(1 + \frac{1}{3}) = 2 + \frac{2}{3} + \frac{3}{4} + \frac{3}{12}$

$= 2 + \frac{8}{12} + \frac{9}{12} + \frac{3}{12}$

$= 2 + \frac{20}{12}$

$= 3\frac{8}{12}$ or $3\frac{2}{3}$

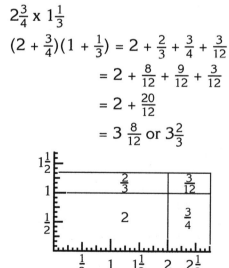

Expanding on Fractions provides the opportunity to practice this algorithm and construct the representations.

Expanding on Fractions

Please show the multiplication of fractions in two ways as shown in the example:
   a. using the distributive property and expanded notation
   b. graphically displaying the distributive property process

1.

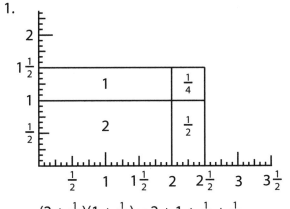

$$(2 + \tfrac{1}{2})(1 + \tfrac{1}{2}) = 2 + 1 + \tfrac{1}{2} + \tfrac{1}{4}$$
$$= 3 + \tfrac{2}{4} + \tfrac{1}{4} \qquad = 3\tfrac{3}{4}$$

2.

$$2\left(1 + \tfrac{3}{4}\right) =$$
$$= \qquad\qquad =$$

3.

$$\left(2 + \tfrac{3}{4}\right)\left(1 + \tfrac{1}{3}\right) =$$
$$= \qquad\qquad =$$

4.

$$\tfrac{2}{3}\left(1 + \tfrac{1}{2}\right) =$$
$$= \qquad\qquad =$$

5.

$$\left(1 + \tfrac{3}{4}\right)\left(1 + \tfrac{3}{4}\right) =$$
$$= \qquad\qquad =$$

6.

$$\left(2 + \tfrac{2}{3}\right)\left(1 + \tfrac{1}{4}\right) =$$
$$= \qquad\qquad =$$

Expanding on Fractions

7.

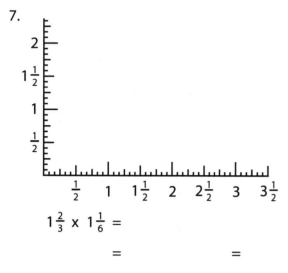

$1\frac{2}{3}$ x $1\frac{1}{6}$ =

=   =

8.

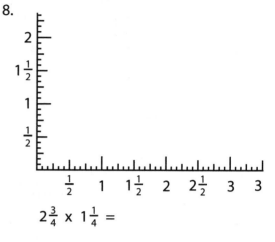

$2\frac{3}{4}$ x $1\frac{1}{4}$ =

=   =

9.

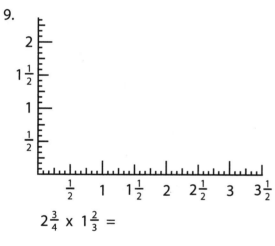

$2\frac{3}{4}$ x $1\frac{2}{3}$ =

=   =

10.

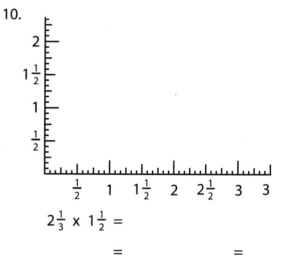

$2\frac{1}{3}$ x $1\frac{1}{2}$ =

=   =

11.

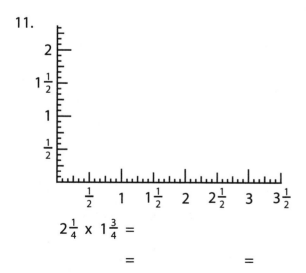

$2\frac{1}{4}$ x $1\frac{3}{4}$ =

=   =

12.

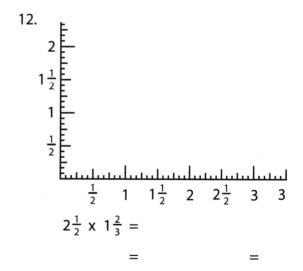

$2\frac{1}{2}$ x $1\frac{2}{3}$ =

=   =

Solutions

Expanding on Fractions

1.
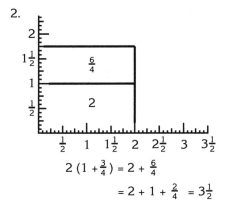

$(2 + \frac{1}{2})(1 + \frac{1}{2}) = 2 + 1 + \frac{1}{2} + \frac{1}{4}$

$= 3 + \frac{2}{4} + \frac{1}{4} = 3\frac{3}{4}$

2.

$2(1 + \frac{3}{4}) = 2 + \frac{6}{4}$

$= 2 + 1 + \frac{2}{4} = 3\frac{1}{2}$

3.
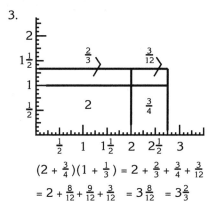

$(2 + \frac{3}{4})(1 + \frac{1}{3}) = 2 + \frac{2}{3} + \frac{3}{4} + \frac{3}{12}$

$= 2 + \frac{8}{12} + \frac{9}{12} + \frac{3}{12} = 3\frac{8}{12} = 3\frac{2}{3}$

4.
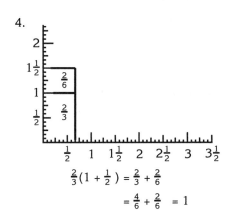

$\frac{2}{3}(1 + \frac{1}{2}) = \frac{2}{3} + \frac{2}{6}$

$= \frac{4}{6} + \frac{2}{6} = 1$

5.
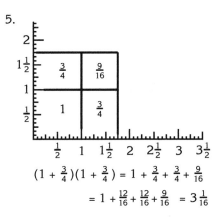

$(1 + \frac{3}{4})(1 + \frac{3}{4}) = 1 + \frac{3}{4} + \frac{3}{4} + \frac{9}{16}$

$= 1 + \frac{12}{16} + \frac{12}{16} + \frac{9}{16} = 3\frac{1}{16}$

6.
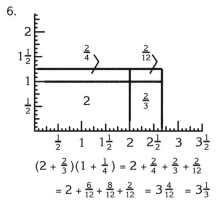

$(2 + \frac{2}{3})(1 + \frac{1}{4}) = 2 + \frac{2}{4} + \frac{2}{3} + \frac{2}{12}$

$= 2 + \frac{6}{12} + \frac{8}{12} + \frac{2}{12} = 3\frac{4}{12} = 3\frac{1}{3}$

7.
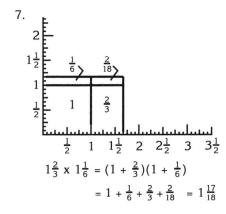

$1\frac{2}{3} \times 1\frac{1}{6} = (1 + \frac{2}{3})(1 + \frac{1}{6})$

$= 1 + \frac{1}{6} + \frac{2}{3} + \frac{2}{18} = 1\frac{17}{18}$

8.
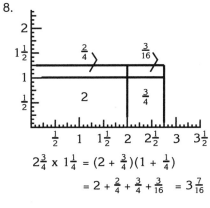

$2\frac{3}{4} \times 1\frac{1}{4} = (2 + \frac{3}{4})(1 + \frac{1}{4})$

$= 2 + \frac{2}{4} + \frac{3}{4} + \frac{3}{16} = 3\frac{7}{16}$

9.
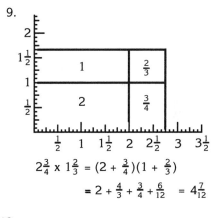

$2\frac{3}{4} \times 1\frac{2}{3} = (2 + \frac{3}{4})(1 + \frac{2}{3})$

$= 2 + \frac{4}{3} + \frac{3}{4} + \frac{6}{12} = 4\frac{7}{12}$

10.
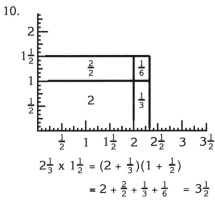

$2\frac{1}{3} \times 1\frac{1}{2} = (2 + \frac{1}{3})(1 + \frac{1}{2})$

$= 2 + \frac{2}{2} + \frac{1}{3} + \frac{1}{6} = 3\frac{1}{2}$

11.
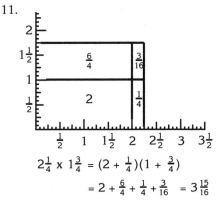

$2\frac{1}{4} \times 1\frac{3}{4} = (2 + \frac{1}{4})(1 + \frac{3}{4})$

$= 2 + \frac{6}{4} + \frac{1}{4} + \frac{3}{16} = 3\frac{15}{16}$

12.
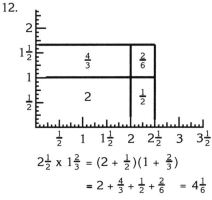

$2\frac{1}{2} \times 1\frac{2}{3} = (2 + \frac{1}{2})(1 + \frac{2}{3})$

$= 2 + \frac{4}{3} + \frac{1}{2} + \frac{2}{6} = 4\frac{1}{6}$

Transitioning to Other Number Bases

lgebra is often defined as generalized arithmetic. This suggests a sequence from the specific to the general. The activities in this volume follow this sequence. Students have already worked with the essential algebraic concepts and processes while still in the familiar environment of the base-ten numeration system. Now they will transfer those same concepts and processes to what can be thought of as other number bases. In fact, equations in some of the activities will be expressed in both numerical and parallel literal notation. This marks the transitional mid-way point on the way to the use of literal notation. Convinced of the general applicability of these concepts and processes in these activities, students will be more comfortable when literal notation is introduced.

Number bases as used here must be understood as being hybrid in nature. While terms that involve x behave like those in various number bases, the numerical value of the expression is still computed in terms of base ten. Nevertheless, important elements of number base numeration are incorporated into the expressions. When viewed from the perspective of number bases, the value of the unknown in any equation with one unknown is the number base in which it is written. This perspective adds interest to the activities in *Part Five*. Students who have mastered the concepts introduced in preceding sections have built the foundation for making the transition to number bases other than ten and then moving on to the use of algebraic expressions.

Students continue to use their knowledge of base-ten arithmetic facts and computation processes. This enables them to check their work and build their confidence. The experiences illustrate that the power of algebra is revealed through its ability to generalize over all number bases.

Style Tiles

Working in a simulated real-world situation, students are challenged to help a company develop the design of a new floor tile. They have three shapes available for the tile design: a large square, a trim strip, and a tab (small square). After completing a prototype, students are required to determine the dimensions, perimeter, and area.

Filling Par 3, 4, and 5 Frames

In this series of three parallel activities, students find the length x width = area quadratic equations in bases three, four, and five. By expressing each relationship in terms of both the base and x, they come to understand the role of literal notation as a way of generalizing the situation. In the process, students use exponential notations such as 3^2, 4^2, 5^2, and x^2 and collect terms.

Par 3, 4, and 5 Constructions

In each instance, students are given a specific set of flats, sticks, and units for building into a rectangle. They are asked to draw a sketch of the solution and express the dimensions and area in terms of a number base and x. In the process, they collect like terms. All constructions are in the first quadrant.

Mystery Numbers

Two types of information are provided: the tile layout of a rectangle and the numerical value of the area expressed in base ten. Students use estimation and trial and error to decipher which number base tiles are involved. This activity provides a unique approach to building understanding of expressions using literal notation.

Style Tiles

Topic
Measurement: perimeter, area

Key Question
How can you determine the perimeter and area of a tile once the dimensions of the base component are known?

Learning Goals
Students will:
- deepen their understanding of the meaning and purpose of perimeter and area,
- accurately calculate perimeters and areas of rectangular regions, and
- generalize measurements of a figure algebraically when one dimension is unknown.

Guiding Documents
Project 2061 Benchmarks
- *Calculate the circumferences and areas of rectangles, triangles, and circles, and the volumes of rectangular solids.*
- *Organize information in simple tables and graphs and identify relationships they reveal.*

*Common Core State Standards for Mathematics**
- *Reason abstractly and quantitatively. (MP.2)*
- *Construct viable arguments and critique the reasoning of others. (MP.3)*
- *Model with mathematics. (MP.4)*
- *Look for and make use of structure. (MP.7)*
- *Work with radicals and integer exponents. (8.EE.A)*

Math
Measurement
    perimeter
    area
Generalizing with expressions
Simplifying expressions

Integrated Processes
Observing
Comparing and contrasting
Collecting and recording data
Generalizing

Materials
AIMS base tiles (see *Management 3*)
Student sheets

Background Information
The scenario in this activity is of a company using a prototype to develop a new tile product. The product is made from sheets of dense foam. The foam is cut into three different components that are assembled in different arrangements and are fused together to form a single tile. The base component is a large square of foam. The size of this base square is the debate within the company. As a prototype it would be best to consider the length as n, making the base component's dimensions $(n \times n)$. A second component is a trim strip. The trim strip's width is one unit, and it will be cut as long as the base component's edge. The strip's dimensions are $(1 \times n)$. The third component is the tab. It is a square one unit on each side (1×1).

Many students have difficulty with the concepts of perimeter and area because they have not internalized what type of measurements they are making. The scenario used in this activity has students visualize the measurement being made. They can imagine that tape will go around the perimeter of the components to hold them together. The amount of floor covered by the tile is the area. By repeatedly calculating these measurements, students gain a familiarity with the measurements and the processes for calculating the measurements. Being familiar with a method of calculation provides the understanding of generalizing these measurements in algebraic expressions.

The different methods used by students to calculate a measurement give rise to different algebraic expressions. The study of these different expressions provides a rich experience for developing literacy in algebraic communication.

Consider the following style example:

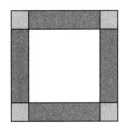

To determine the perimeter, many students might start on a corner and go around adding pieces ($1 + n + 2 + n + 2 + n + 2 + n + 1$). Students are quick to see that this can be written much more efficiently by combining the numbers and the variables ($4n + 8$). Other students will approach the perimeter by recognizing there are four equal sides with the dimension ($n + 2$). They will express the perimeter as $4(n + 2)$. This provides an opportunity to practice the distributive property to find the equivalence of $4(n + 2) = (4n + 8)$. When determining the area of the style, most students will count up each type of component; one base, four strips, and four tabs would be expressed ($n^2 + 4n + 4$). This provides an opportunity to see whether the formula of length times width equals area works algebraically. Students can be asked to determine the length and width of the style ($n + 2$) and then substitute it in the formula to see if it produces the equivalent expression $(n + 2)(n + 2) = (n^2 + 2n + 2n + 4) = (n^2 + 4n + 4)$.

Management

1. How students approach the situation of dealing with a prototype allows the teacher to assess what understanding of a variable a student has and gives an idea of the level of abstraction at which a student is working. Students who work out the measurement for base components of 3, 4, 5, and 8 units and then struggle to write an algebraic expression are just beginning to understand the meaning of a variable and are strongly concrete thinkers. Students who want to work with the algebraic expression and then substitute in the values of 3, 4, 5, and 8 for n are looking at the situation with abstract understanding and have developed an understanding of variable. Adjust the approach to best meet the needs of most of the students in the class.
2. The time required to complete this activity varies greatly with the experience students have had with measurement and the level of abstraction at which they are working.
3. Students will need a set of AIMS base tiles for this activity. The tiles can be base ten, or any other base.
4. One set of tiles for each group of four students is optimal. Working in small groups allows students to share understanding and clarify problems.

Procedure

1. Distribute the tiles and student sheets and discuss the scenario. Some time may need to be taken to make sure that students understand the concept of a prototype and that the dimensions of the base unit (n) are not determined.
2. Direct students to determine the perimeter and area of each type of tile when the base component is 3, 4, 5, and 8 units long. (See *Solutions*.)

3. Have students discuss and algebraically determine the dimensions and area of each type of tile when the base component is n units long. (See *Solutions*.)
4. Using the tiles, have the students construct each of the tile styles.
5. Direct the students to determine what the perimeter and area would be for each style if the base unit's dimension were changed. For the algebraic expression, you may need to spend some time on combining like terms or using the distributive property to confirm that expressions are equivalent. Refer to *Background Information* for more details on this process.

Connecting Learning

1. How did you find the perimeter and area for each style? [Various. Counted the lengths around the edges, counted the squares strips and tabs, multiplied the length by the width, etc.]
2. How did finding the measurements for different base lengths help you generate the algebraic expressions in terms of n?
3. How could you get the expression for the area of a style without counting the pieces? [Record the length and width in terms of n, and then multiply them using the distributive property.]
4. Why is the algebraic expression using n units the most useful in dealing with a prototype? [Once the base dimension is determined, it can be substituted into the expression and the calculation only needs to be done once.]

Extensions

1. Have students construct their own tile styles, record pictures of them, and determine the algebraic expressions of the perimeter and area of the styles. Direct the students to exchange their pictures and see if other students can determine the algebraic expressions for the measurements.
2. Have students record the length and width of different styles in terms of n and then use them in the formulas $l \times w = A$ and $2(l + w) = P$ and see if they can generate the perimeter and area expressions for the each style.
3. Have students complete *Tile Styles Rearranged* to help them relate the measurement process to the algebraic concept of factors and products.

Solutions

The tables that appear on the student sheets are given here with the correct answers filled in.

Style Tiles Pieces

		PROPOSED DIMENSION OF BASE COMPONENT				
		3 units	4 units	5 units	8 units	n units
Perimeter	Tab	4	4	4	4	4
	Strip	8	10	12	18	$2n + 2$
	Base	12	16	20	32	$4n$

Area	Tab	1	1	1	1	1
	Strip	3	4	5	8	n
	Base	9	16	25	64	n^2

Style Tiles Measurements

		PROPOSED DIMENSION OF BASE COMPONENT				
		3 units	4 units	5 units	8 units	n units
Perimeter STYLES	A	16	20	24	36	$4n + 4$
	B	20	24	28	40	$4n + 8$
	C	18	22	26	38	$4n + 6$
	D	28	32	36	48	$4n + 16$
	E	26	32	38	56	$6n + 8$
	F	32	40	48	72	$8n + 8$
	G	28	36	44	68	$8n + 4$
	H	36	44	52	76	$8n + 12$

		PROPOSED DIMENSION OF BASE COMPONENT				
		3 units	4 units	5 units	8 units	n units
Area STYLES	A	15	24	35	80	$n^2 + 2n$
	B	25	36	49	100	$n^2 + 4n + 4$
	C	20	30	42	90	$n^2 + 3n + 2$
	D	49	64	81	144	$n^2 + 8n + 16$
	E	36	55	78	171	$2n^2 + 5n + 3$
	F	64	100	144	324	$4n^2 + 8n + 4$
	G	49	81	121	289	$4n^2 + 4n + 1$
	H	81	121	169	361	$4n^2 + 12n + 9$

Style Tiles Rearranged (Extension 3)

See the tables for the areas.

B.

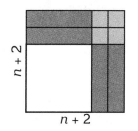

C.

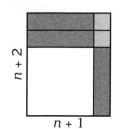

D.

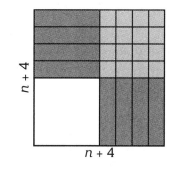

E.

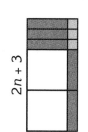

F.

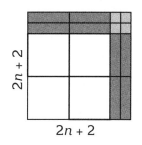

G.

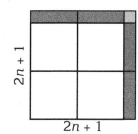

H.

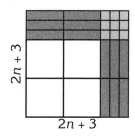

You work for a company that is developing a new floor tile that will be marketed around the country. It is still in its prototype stage and management has come to you for your expertise in measurement.

The tile is made from a dense foam. There are three different components that are assembled in different arrangements and are fused together to form a single tile. The base component is a large square of foam. A second component is a trim strip. The trim strip's width is one unit and it will be cut as long as the base component's edge. The third component is the tab. It is a square with edges of one unit to match the trim strip.

The company has yet to determine the size of the basic component. The engineering department has made some proposals, but the marketing department will decide what will sell best. They have given you the prototype material to use to help you in your calculations.

Style Tiles

The company will use the following types of pieces to make floor tiles. Use your prototype materials and determine the measurement of the tiles shown here for each of the dimensions given for the base components length. Record your data in the tables below.

Tab

Trim Strip

Base Component

	PROPOSED DIMENSION OF BASE COMPONENT				
		3 units	4 units	5 units	8 units	*n* units
Perimeter	Tab					
	Strip					
	Base					
Area	Tab					
	Strip					
	Base					

Style Tiles Measurements

The company is considering the following styles of floor tiles. Construct them from your prototype materials and complete the tables on the following page with their perimeters and areas as requested.

A.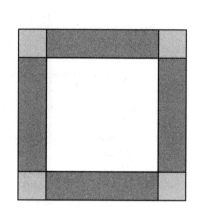

B.

C.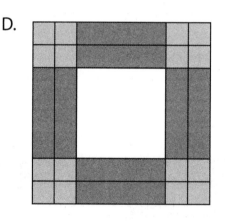

D.

E.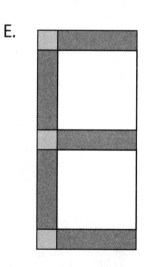

F.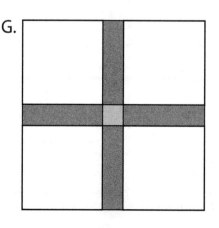

G.

H.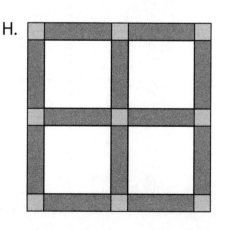

Style Tiles

Fusing tape surrounds the arrangement of components to create a single tile. The company has not yet determined what size to make the base component's length. Calculate the amount of tape needed for the various styles with each proposed base length.

	PROPOSED DIMENSION OF BASE COMPONENT				
Perimeter STYLES	3 units	4 units	5 units	8 units	n units
A					
B					
C					
D					
E					
F					
G					
H					

Interior designers need to know how much area each tile will cover. The company has not yet determined what size to make the base component's length. Calculate the area covered by the various styles for each proposed base length.

	PROPOSED DIMENSION OF BASE COMPONENT				
Area STYLES	3 units	4 units	5 units	8 units	n units
A					
B					
C					
D					
E					
F					
G					
H					

Style Tiles *Rearranged*

Rearrange the components of each style to fit in the same area by placing all the base components in the lower left corner, all the tabs in the upper right corner, and all the trim strips along the top and the right sides. Draw a picture of the rearranged tile and record the dimensions and area for each style assuming that the base component is n. The first problem has been done for you as an example.

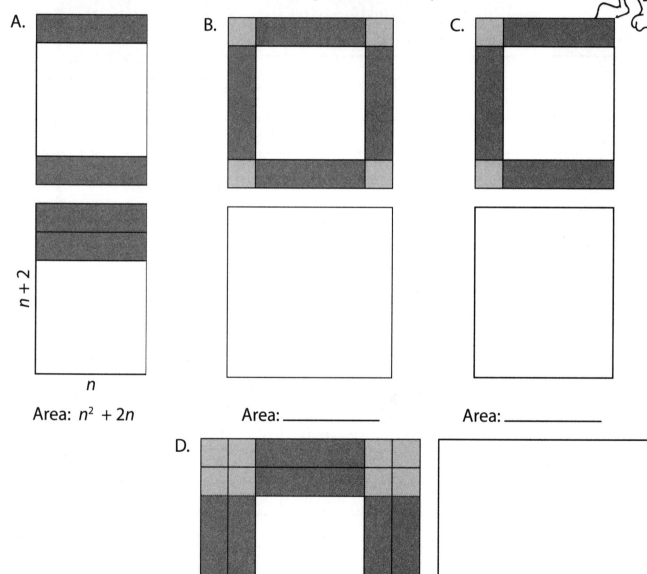

A.

Area: $n^2 + 2n$

B.

Area: _____

C.

Area: _____

D.

Area: _____

Using the back of this paper, describe how the rearrangement displays the dimensions and areas more readily.

E.

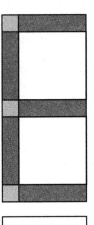

Area: _____

F.

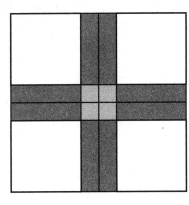

Area: _____

G.

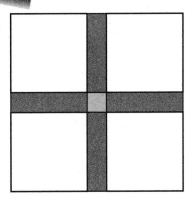

Area: _____

H.

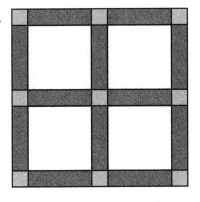

Area: _____

Connecting Learning

1. How did you find the perimeter and area for each style?

2. How did finding the measurements for different base lengths help you generate the algebraic expressions in terms of *n*?

3. How could you get the expression for the area of a style without counting the pieces?

4. Why is the algebraic expression using *n* units the most useful in dealing with a prototype?

PAR 3, 4, & 5

Topic
Distributive property

Key Question
How can you find the length, width, and area of a rectangle using base three (or four or five) tiles?

Learning Goals
Students will:
- arrange sets of base three (or four or five) tiles into a rectangle, and
- write a mathematical sentence for length x width = area in base three (or four or five) using expanded and exponential notation.

Guiding Document
Common Core State Standards for Mathematics *
- *Model with mathematics. (MP.4)*
- *Use appropriate tools strategically. (MP.5)*
- *Look for and make use of structure. (MP.7)*
- *Use place value understanding and properties of operations to perform multi-digit arithmetic. (4.NBT.B)*

Math
Distributive property
Multiplication
Expanded notation
Base three, four, and five
Algebraic notation

Integrated Processes
Observing
Collecting and recording data
Comparing and contrasting
Generalizing

Materials
Base three, four, and five tiles
Rulers, optional
Student sheets

Background Information
This marks the introduction of number bases other than ten. In this series of three parallel activities, students will find the length x width = area quadratic equations in bases three, four, and five using arithmetic numbers with the same set of outlines. By expressing each relationship in terms of both the base and x, they will come to understand the algebraic role of x as a variable. At one time x represents three, then four, and then five or any other number.

The use of number bases is crucial in building a bridge between arithmetic and algebra. It leads students to understand the role of variables while still largely in the realm of arithmetic numbers.

Students will use exponential notations such as 3^2, 4^2, 5^2, and x^2. If they have not learned such notation, it should be taught along with the rest of this lesson.

In each case, the dimensions of the outline define the factors in the quadratic equation and the area defines the product.

Management
1. Students will need to have base three, four, and five tiles to do this activity. One set of each can be shared between a few students.
2. Be sure that students have done some work with base-ten tiles before they do this activity. Without the background of working on similar problems in a familiar base, students may have trouble making the jump to an unfamiliar base.
3. For each number base there is a part one and a part two. The second part is included for practice, and is not necessary if students have grasped the concept well after the first sheet.

Procedure
1. Hand out base-three tiles and the first two student sheets to each student and go over the instructions. Be sure that students understand how to do each of the sections, especially the transfer of the variables into algebraic notation.
2. Have students work independently or in small groups to complete the first student sheet.
3. When all groups have finished, hand out the *Filling Par Four Frames* student sheet and the base-four tiles.
4. When students have completed the information using base four, hand out the student sheet and tiles so that they can do the same for base five.
5. If desired, hand out the second set of frames and have students complete the sheets for *Filling Par Frames 2.*
6. After all three sections have been completed, close with a time of class discussion and sharing where students make comparisons between the different bases and share what they have learned.

Connecting Learning

1. How is this activity similar to others that you have done before? [The process is the same as that for any activity that uses base-ten tiles.] How is it different? [The number base is different.]
2. What things changed as you went from base three to base four to base five? [the number of flats, longs, and units that would fit in each square]
3. What things stayed the same? [the area of the rectangle]
4. What would the algebraic notation be for the fourth rectangle in base eight? [$x^2 + 2x + 0 = 80$] How do you know?
5. What did this activity teach you about algebraic notation?
6. What other things did you learn while doing this activity?

Extension

Challenge more advanced students to factor their algebraic expressions from part (d) and compare the results from base three to base four to base five.

Solutions

The solutions for both sections of *Filling Par Frames* are shown without the illustrations.

Par Three

1.
b.   1 flat + 3 longs + 2 units
c. $3^2 + 3(3) + 2 = 20$
d. $x^2 + 3x + 2 = 20$

3.
b.   6 flats + 3 longs + 0 units
c. $6(3^2) + 3(3) + 0 = 63$
d. $6x^2 + 3x = 63$

2.
b.   2 flats + 5 longs + 2 units
c. $2(3^2) + 5(3) + 2 = 35$
d. $2x^2 + 5x + 2 = 35$

5.
b.   2 flats + 4 longs + 0 units
c. $2(3^2) + 4(3) + 0 = 30$
d. $2x^2 + 4x = 30$

4.
b.   6 flats + 8 longs + 2 units
c. $6(3^2) + 8(3) + 2$ units = 80
d. $6x^2 + 8x + 2$ units = 80

7.
b.   4 flats + 6 longs + 2 units
c. $4(3^2) + 6(3) + 2 = 56$
d. $4x^2 + 6x + 2 = 56$

6.
b.   4 flats + 2 longs + 0 units
c. $4(3^2) + 2(3) + 0 = 42$
d. $4x^2 + 2x = 42$

8.
b.   6 flats + 6 longs + 0 units
c. $6(3^2) + 6(3) + 0 = 7$
d. $6x^2 + 6x = 72$

Par Four

1.
b.   1 flat + 1 long + 0 units
c. $4^2 + 4 + 0 = 20$
d. $x^2 + x = 20$

2.
b.   1 flat + 4 longs + 3 units
c. $4^2 + 4(4) + 3 = 35$
d. $x^2 + 4x + 3 = 35$

3.
b.   2 flats + 7 longs + 3 units
c. $2(4^2) + 7(4) + 3 = 63$
d. $2x^2 + 7x + 3 = 63$

4.
b.   4 flats + 4 longs + 0 units
c. $4(4^2) + 4(4) + 0 = 80$
d. $4x^2 + 4x = 80$

5.
b.   1 flat + 3 longs + 2 units
c. $4^2 + 3(4) + 2 = 30$
d. $x^2 + 3x + 2 = 30$

6.
b.   1 flat + 5 longs + 6 units
c. $4^2 + 5(4) + 6 = 42$
d. $x^2 + 5x + 6 = 42$

7.
b.   2 flats + 6 longs + 0 units
c. $2(4^2) + 6(4) + 0 = 56$
d. $2x^2 + 6x = 56$

8.
b.   4 flats + 2 longs + 0 units
c. $4(4^2) + 2(4) + 0 = 72$
d. $4x^2 + 2x = 72$

Par Five

1.
b.   0 flats + 4 longs + 0 units
c. $0 + 4(5) + 0 = 20$
d. $4x = 20$

2.
b.   1 flat + 2 longs + 0 units
c. $5^2 + 2(5) + 0 = 35$
d. $x^2 + 2x = 35$

3.
b.   1 flat + 6 longs + 8 units
c. $5^2 + 6(5) + 8 = 63$
d. $x^2 + 6x + 8 = 63$

4.
b.   2 flats + 6 longs + 0 units
c. $2(5^2) + 6(5) + 0 = 80$
d. $2x^2 + 6x = 80$

5.
b.   1 flat + 1 long + 0 units
c. $5^2 + 5 + 0 = 30$
d. $x^2 + x = 30$

6.
b.   1 flat + 3 longs + 2 units
c. $5^2 + 3(5) + 2 = 42$
d. $x^2 + 3x + 2 = 42$

7.
b.   1 flat + 5 longs + 6 units
c. $5^2 + 5(5) + 6 = 56$
d. $x^2 + 5x + 6 = 56$

8.
b.   1 flat + 7 longs + 12 units
c. $5^2 + 7(5) + 12 = 72$
d. $x^2 + 7x + 12 = 72$

FILLING
PAR 3, 4, & 5
FRAMES

1

2

3

4

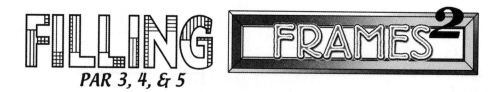

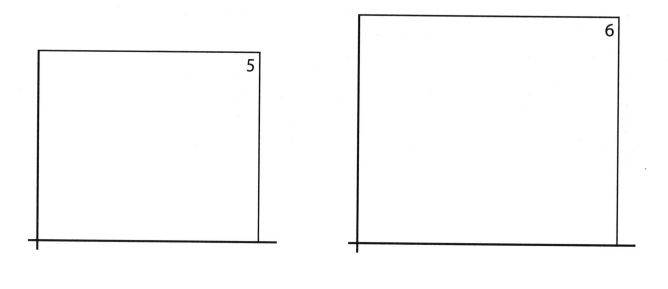

5

6

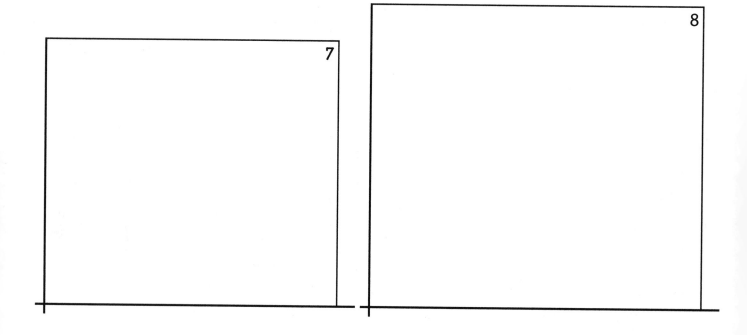

7

8

Please cover the rectangles with the fewest possible base three flats, longs, and units. Begin at the origin with the flats, making sure each flat is as close to the origin as possible. Then build outward using lengths next and units last. All horizontal and vertical breaks between pieces must extend through the entire rectangle.

In part a, sketch a picture of each solution using squares, lines and dots to represent the flats, longs, and units. In part b, record the number of flats, longs, and units used as an addition sentence. Then, write a mathematical sentence in terms of the base using exponential notation in part c. Finally, in part d, substitute x for the number base and translate the equation into algebraic notation. One problem has been done for you as an example.

1a.

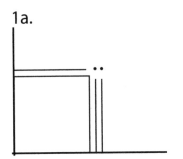

b. _1_ flat + _3_ longs + _2_ units
c. $(3^2) + 3(3) + 2 = 20$
d. $x^2 + 3x + 2 = 20$

2a.

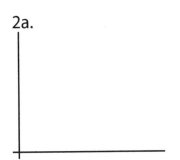

b. ____ flats + ____ longs + ____ units

c. $(3^2)+$ $(3) +$ $=$

d. $x^2 +$ $x +$ $=$

3a.

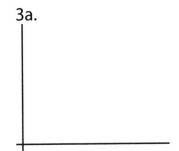

b. ____ flats + ____ longs + ____ units

c. $(3^2)+$ $(3) +$ $=$

d. $x^2 +$ $x +$ $=$

4a.

b. ____ flats + ____ longs + ____ units

c. $(3^2)+$ $(3) +$ $=$

d. $x^2 +$ $x +$ $=$

PAR THREE

Repeat the process from *Filling Par Three Frames 1* with the second set of frames.

5a.

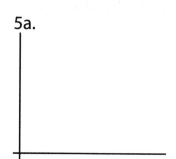

b. _____ flats + _____ longs + _____ units

c. _____ (3^2)+ _____ (3) + _____ = _____

d. _____ x^2 + _____ x + _____ = _____

6a.

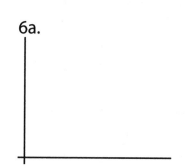

b. _____ flats + _____ longs + _____ units

c. _____ (3^2)+ _____ (3) + _____ = _____

d. _____ x^2 + _____ x + _____ = _____

7a.

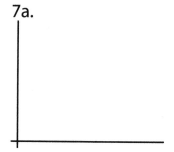

b. _____ flats + _____ longs + _____ units

c. _____ (3^2)+ _____ (3) + _____ = _____

d. _____ x^2 + _____ x + _____ = _____

8a.

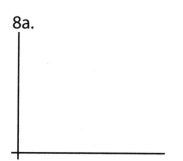

b. _____ flats + _____ longs + _____ units

c. _____ (3^2)+ _____ (3) + _____ = _____

d. _____ x^2 + _____ x + _____ = _____

FILLING FRAMES 1
PAR FOUR

Please repeat the process you used in *Filling Par Three Frames* with base four flats, longs, and units. You should use the same rectangles, and write the same sentences in terms of base four.

1a.

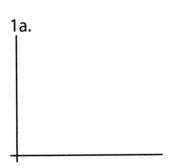

b. ____ flats + ____ longs + ____ units

c. $(4^2)+$ $(4) +$ $=$

d. $x^2 +$ $x +$ $=$

2a.

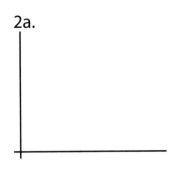

b. ____ flats + ____ longs + ____ units

c. $(4^2)+$ $(4) +$ $=$

d. $x^2 +$ $x +$ $=$

3a.

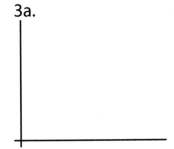

b. ____ flats + ____ longs + ____ units

c. $(4^2)+$ $(4) +$ $=$

d. $x^2 +$ $x +$ $=$

4a.

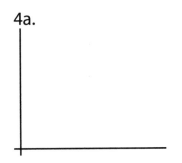

b. ____ flats + ____ longs + ____ units

c. $(4^2)+$ $(4) +$ $=$

d. $x^2 +$ $x +$ $=$

FILLING
PAR FOUR

Repeat the process from *Filling Par Four Frames 1* with the second set of frames.

5a.

b. _____ flats + _____ longs + _____ units

c. $(4^2)+$ $(4) +$ $=$

d. $x^2 +$ $x +$ $=$

6a.

b. _____ flats + _____ longs + _____ units

c. $(4^2)+$ $(4) +$ $=$

d. $x^2 +$ $x +$ $=$

7a.

b. _____ flats + _____ longs + _____ units

c. $(4^2)+$ $(4) +$ $=$

d. $x^2 +$ $x +$ $=$

8a.

b. _____ flats + _____ longs + _____ units

c. $(4^2)+$ $(4) +$ $=$

d. $x^2 +$ $x +$ $=$

PAR FIVE

Please repeat the process you used in *Filling Par Three Frames* and *Filling Par Four Frames* with base five flats, longs, and units. You should use the same rectangles, and write the same sentences in terms of base five.

1a.

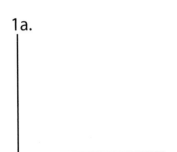

b. ____ flats + ____ longs + ____ units

c. $(5^2)+$ $(5) +$ =

d. $x^2 +$ $x +$ =

2a.

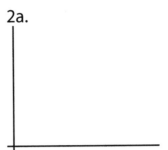

b. ____ flats + ____ longs + ____ units

c. $(5^2)+$ $(5) +$ =

d. $x^2 +$ $x +$ =

3a.

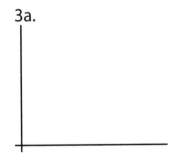

b. ____ flats + ____ longs + ____ units

c. $(5^2)+$ $(5) +$ =

d. $x^2 +$ $x +$ =

4a.

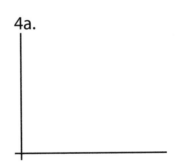

b. ____ flats + ____ longs + ____ units

c. $(5^2)+$ $(5) +$ =

d. $x^2 +$ $x +$ =

FILLING FRAMES²
PAR FIVE

Repeat the process from *Filling Par Five Frames 1* with the second set of frames.

5a.

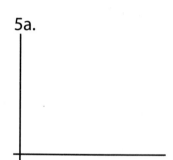

b. _____ flats + _____ longs + _____ units

c. $(5^2)+$ $(5) +$ $=$

d. $x^2 +$ $x +$ $=$

6a.

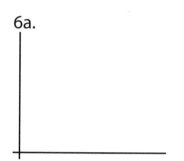

b. _____ flats + _____ longs + _____ units

c. $(5^2)+$ $(5) +$ $=$

d. $x^2 +$ $x +$ $=$

7a.

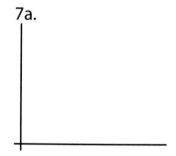

b. _____ flats + _____ longs + _____ units

c. $(5^2)+$ $(5) +$ $=$

d. $x^2 +$ $x +$ $=$

8a.

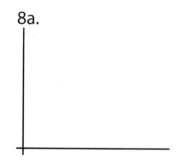

b. _____ flats + _____ longs + _____ units

c. $(5^2)+$ $(5) +$ $=$

d. $x^2 +$ $x +$ $=$

When you have completed the student sheets for bases three, four, and five, answer the following questions.

1. How is this activity similar to others that you have done before? How is it different?

2. What things changed as you went from base three to base four to base five?

3. What things stayed the same?

4. What would the algebraic notation be for the fourth rectangle in base eight? How do you know?

5. What did this activity teach you about algebraic notation?

6. What other things did you learn while doing this activity?

Connecting Learning

1. How is this activity similar to others that you have done before? How is it different?

2. What things changed as you went from base three to base four to base five?

3. What things stayed the same?

4. What would the algebraic notation be for the fourth rectangle in base eight? How do you know?

5. What did this activity teach you about algebraic notation?

6. What other things did you learn while doing this activity?

PAR 3 4 and 5 Constructions

Topic
Distributive property

Key Question
How can a set of base three, four, or five tiles be arranged to form a rectangle?

Learning Goals
Students will:
- arrange sets of base three, four, or five tiles to form rectangles in the first quadrant;
- determine the dimensions in both base three, four or five, and x;
- use the distributive property to compute the area and complete the length x width = area mathematical sentence; and
- check the results of their computation of the area against the tiles in the rectangle.

Guiding Documents
Project 2061 Benchmark
- *Mathematical ideas can be represented concretely, graphically, and symbolically.*

Common Core State Standards for Mathematics*
- *Reason abstractly and quantitatively. (MP.2)*
- *Model with mathematics. (MP.4)*
- *Look for and make use of structure. (MP.7)*
- *Work with radicals and integer exponents. (8.EE.A)*

Math
Distributive property
Multiplication
Expanded notation
Bases three, four, five, and x

Integrated Processes
Observing
Collecting and recording data
Comparing
Generalizing

Materials
Base three, four, or five tiles
Rulers, optional
Student sheets

Background Information
In this activity, students will begin with a prescribed set of base three, four, or five tiles and piece them together to form a rectangle following standard rules. The resulting arrangement defines the length and width dimensions. Students will record the value both in the numerical base and base x. Next, they will apply the distributive property in multiplying the length by width to obtain the area of the rectangle. Finally, they will check the results from multiplying with the layout of the rectangle.

To prepare for this activity, students should be taught algebraic products such as x times $x = x^2$ and 3 times $x = 3x$.

Caution: Students will be tempted to complete the mathematical sentences by looking at the picture rather than using the distributive property. Precautions will need to be taken to prevent this since the purpose is to exercise the use of the distributive property. Students might, for example, be asked to cover the original description of the set and the sketch once the dimensions are recorded and treat it as a mystery to be solved by applying the distributive property. The reward comes from seeing that the results of the computation match the distribution of tiles in the sketch.

Have students follow the same use of not writing the coefficient 1. For example, in the first exercise in *Par Three Constructions* the coefficient for both the (3^2) term and the two (3) terms is 1 before the terms are collected. Have students simply record the answer as $3^2 + 3 + 3 + 1 = 3^2 + 2(3) + 1$.

In exercise 3 and following, the 3^2 has a coefficient other than 1. Since parentheses are used to separate length and width, students will need to be introduced to using the dot as a multiplication sign. Two times 3^2 will be shown as $2 \cdot 3^2$.

Management
1. Students will each need the appropriate set of tiles to complete this activity. One set can be shared between two or three students.
2. Students should be familiar with how to use the tiles to construct the rectangle, how to determine the dimensions in both the numerical and algebraic base, and how to write an equation in expanded and exponential notation before beginning this activity.

Procedure

1. Hand out the appropriate student sheets and tiles to each student. Go over the instructions as a class and be sure students understand how to organize their flats, longs, and units into the four groups. *It is possible to make a rectangle from each set of pieces listed. Create a rectangle on your desk using those pieces, then make a sketch of it in the space provided and record the length and width in both the numeric and algebraic forms using expanded notation. Cover the description of the set of tiles and the sketch. Use the distributive property to determine the area both arithmetically and algebraically. Compare your results with the initial description of the set and the sketch.*

2. If necessary, do an example together as a class, then have students work together in groups to complete the student sheet. Encourage students to find multiple solutions to the problems in the cases where that is possible.

3. Once all students have completed the problems, close with a time of class discussion and sharing.

Connecting Learning

1. How was this activity like others that you have done? How was it different?
2. How did the numeric and algebraic forms of the equations compare? [They were the same with the base number represented by the x in the algebraic form.]
3. How did the results of your computations compare with the sketches you made of each set of tiles? [The number of flats, longs, and units are represented by the numbers in the equations.]
4. What changed about your equations when there was more than one flat?
5. What did you learn about the distributive property from this activity?

Extension

Have students determine the solutions to similar problems in base six or base seven without using any physical manipulatives.

Solutions

The solutions for all of the student sheets are given without the sketches.

Par Three Constructions

1. 1 flat, 2 longs, 1 unit
   b. $(3 + 1)(3 + 1) =$
      $3^2 + 1(3) + 1(3) + 1 =$
      $3^2 + 2(3) + 1$
   c. $(x + 1)(x + 1) =$
      $x^2 + 1x + 1x + 1 =$
      $x^2 + 2x + 1$

2. 1 flat, 3 longs, 2 units
   b. $(3 + 2)(3 + 1) =$
      $3^2 + 1(3) + 2(3) + 2 =$
      $3^2 + 3(3) + 2$
   c. $(x + 2)(x + 1) =$
      $x^2 + 1x + 2x + 2 =$
      $x^2 + 3x + 2$

3. 2 flats, 5 longs, 2 units
   b. $(2 \cdot 3 + 1)(3 + 2) =$
      $2(3^2) + 4(3) + 1(3) + 2 =$
      $2(3^2) + 5(3) + 2$
   c. $(2x + 1)(x + 1) =$
      $2x^2 + 4x + 1x + 2 =$
      $2x^2 + 5x + 2$

4. 2 flats, 6 longs, 4 units
   b. $(2 \cdot 3 + 2)(3 + 2) =$
      $2(3^2) + 4(3) + 2(3) + 4 =$
      $2(3^2) + 6(3) + 4$
   c. $(2x + 2)(x + 2) =$
      $2x^2 + 4x + 2x + 4 =$
      $2x^2 + 6x + 4$

5. 3 flats, 5 longs, 2 units
   b. $(3 \cdot 3 + 2)(3 + 1) =$
      $3(3^2) + 3(3) + 2(3) + 2 =$
      $3(3^2) + 5(3) + 2$
   c. $(3x + 2)(x + 1) =$
      $3x^2 + 3x + 2x + 2 =$
      $3x^2 + 5x + 2$

6. 3 flats, 7 longs, 2 units
   b. $(3 \cdot 3 + 1)(3 + 2) =$
      $3(3^2) + 6(3) + 1(3) + 2 =$
      $3(3^2) + 7(3) + 2$
   c. $(3x + 1)(x + 2) =$
      $3x^2 + 6x + 1x + 2 =$
      $3x^2 + 7x + 2$

7. 3 flats, 8 longs, 4 units

  b. $(3 \cdot 3 + 2)(3 + 2) =$
$3(3^2) + 6(3) + 2(3) + 4 =$
$3(3^2) + 8(3) + 4$

  c. $(3x + 2)(x + 2) =$
$3x^2 + 6x + 2x + 4 =$
$3x^2 + 8x + 4$

8. 4 flats, 4 longs, 1 unit

  b. $(2 \cdot 3 + 1)(2 \cdot 3 + 1) =$
$4(3^2) + 2(3) + 2(3) + 1 =$
$4(3^2) + 4(3) + 1$

  c. $(2x + 1)(2x + 1) =$
$4x^2 + 2x + 2x + 1 =$
$4x^2 + 4x + 1$

9. 4 flats, 6 longs, 2 units

  b. $(2 \cdot 3 + 2)(2 \cdot 3 + 1) =$
$4(3^2) + 2(3) + 4(3) + 2 =$
$4(3^2) + 6(3) + 2$

  c. $(2x + 2)(2x + 1) =$
$4x^2 + 2x + 4x + 2 =$
$4x^2 + 6x + 2$

Par Four Constructions

1. 1 flat 3 longs, 2 units

  b. $(4 + 2)(4 + 1) =$
$4^2 + 1(4) + 2(4) + 2 =$
$4^2 + 3(4) + 2$

  c. $(x + 2)(x + 1) =$
$x^2 + 1x + 2x + 2 =$
$x^2 + 3x + 2$

2. 1 flat, 6 longs, 9 units

  b. $(4 + 3)(4 + 3) =$
$4^2 + 3(4) + 3(4) + 9 =$
$4^2 + 6(4) + 9$

  c. $(x + 3)(x + 3) =$
$x^2 + 3x + 3x + 9 =$
$x^2 + 6x + 9$

3. 2 flats, 7 longs, 6 units

  b. $(2 \cdot 4 + 3)(4 + 2) =$
$2(4^2) + 4(4) + 3(4) + 6 =$
$2(4^2) + 7(4) + 6$

  c. $(2x + 3)(x + 2) =$
$2x^2 + 4x + 3x + 6 =$
$2x^2 + 7x + 6$

4. 2 flats, 8 longs, 6 units

  b. $(2 \cdot 4 + 2)(4 + 3) =$
$2(4^2) + 6(4) + 2(4) + 6 =$
$2(4^2) + 8(4) + 6$

  c. $(2x + 2)(x + 3) =$
$2x^2 + 6x + 2x + 6 =$
$2x^2 + 8x + 6$

5. 3 flats, 9 longs, 6 units

  b. $(3 \cdot 4 + 3)(4 + 2) =$
$3(4^2) + 6(4) + 3(4) + 6 =$
$3(4^2) + 9(4) + 6$

  c. $(3x + 3)(x + 2) =$
$3x^2 + 6x + 3x + 6 =$
$3x^2 + 9x + 6$

6. 2 flats, 7 longs, 3 units

  b. $(2 \cdot 4 + 1)(4 + 3) =$
$2(4^2) + 6(4) + 1(4) + 3 =$
$2(4^2) + 7(4) + 3$

  c. $(2x + 1)(x + 3) =$
$2x^2 + 6x + 1x + 3 =$
$2x^2 + 7x + 3$

7. 3 flats, 5 longs, 2 units

  b. $(3 \cdot 4 + 2)(4 + 1) =$
$3(4^2) + 3(4) + 2(4) + 2 =$
$3(4^2) + 5(4) + 2$

  c. $(3x + 2)(x + 1) =$
$3x^2 + 3x + 2x + 2 =$
$3x^2 + 5x + 2$

8. 3 flats, 7 longs, 2 units

  b. $(3 \cdot 4 + 1)(4 + 2) =$
$3(4^2) + 6(4) + 1(4) + 2 =$
$3(4^2) + 7(4) + 2$

  c. $(3x + 1)(x + 2) =$
$3x^2 + 6x + 1x + 2 =$
$3x^2 + 7x + 2$

9. 4 flats, 10 longs, 4 units

  b. $(2 \cdot 4 + 4)(2 \cdot 4 + 1) =$
$4(4^2) + 2(4) + 8(4) + 4 =$
$4(4^2) + 10(4) + 4$

  c. $(2x + 4)(2x + 1) =$
$4x^2 + 2x + 8x + 4 =$
$4x^2 + 10x + 4$

Par Five Constructions

1. 1 flat, 6 longs, 9 units

  b. $(5 + 3)(5 + 3) =$
$5^2 + 3(5) + 3(5) + 9 =$
$5^2 + 6(5) + 9$

  c. $(x + 3)(x + 3) =$
$x^2 + 3x + 3x + 9 =$
$x^2 + 6x + 9$

2. 2 flats, 7 longs, 6 units

  b. $(2 \cdot 5 + 3)(5 + 2) =$
$2(5^2) + 4(5) + 3(5) + 6 =$
$2(5^2) + 7(5) + 6$

  c. $(2x + 3)(x + 2) =$
$2x^2 + 4x + 3x + 6 =$
$2x^2 + 7x + 6$

3. 2 flats, 7 longs, 3 units

  b. $(2 \cdot 5 + 1)(5 + 3) =$
$2(5^2) + 6(5) + 1(5) + 3 =$
$2(5^2) + 7(5) + 3$

  c. $(2x + 1)(x + 3) =$
$2x^2 + 6x + 1x + 3 =$
$2x^2 + 7x + 3$

4. 2 flats, 4 longs, 2 units

  b. $(2 \cdot 5 + 2)(5 + 1) =$
$2(5^2) + 2(5) + 2(5) + 2 =$
$2(5^2) + 4(5) + 2$

  c. $(2x + 2)(x + 1) =$
$2x^2 + 2x + 2x + 2 =$
$2x^2 + 4x + 2$

5. 3 flats, 9 longs, 6 units

  b. $(3 \cdot 5 + 3)(5 + 2) =$
$3(5^2) + 6(5) + 3(5) + 6 =$
$3(5^2) + 9(5) + 6$

  c. $(3x + 3)(x + 2) =$
$3x^2 + 6x + 3x + 6 =$
$3x^2 + 9x + 6$

6. 3 flats, 6 longs, 3 units

  b. $(3 \cdot 5 + 3)(5 + 1) =$
$3(5^2) + 3(5) + 3(5) + 3 =$
$3(5^2) + 6(5) + 3$

  c. $(3x + 3)(x + 1) =$
$3x^2 + 3x + 3x + 3 =$
$3x^2 + 6x + 3$

7. 3 flats, 5 longs, 2 units

  b. $(3 \cdot 5 + 2)(5 + 1) =$
$3(5^2) + 3(5) + 2(5) + 2 =$
$3(5^2) + 5(5) + 2$

  c. $(3x + 2)(x + 1) =$
$3x^2 + 3x + 2x + 2 =$
$3x^2 + 5x + 2$

8. 3 flats, 8 longs, 4 units

  b. $(3 \cdot 5 + 2)(5 + 2) =$
$3(5^2) + 6(5) + 2(5) + 4 =$
$3(5^2) + 8(5) + 4$

  c. $(3x + 2)(x + 2) =$
$3x^2 + 6x + 2x + 4 =$
$3x^2 + 8x + 4$

9. 3 flats, 7 longs, 2 units

  b. $(3 \cdot 5 + 1)(5 + 2) =$
$3(5^2) + 6(5) + 1(5) + 2 =$
$3(5^2) + 7(5) + 2$

  c. $(3x + 1)(x + 2) =$
$3x^2 + 6x + 1x + 2 =$
$3x^2 + 7x + 2$

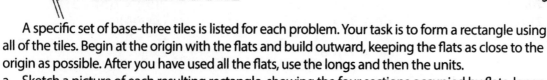

A specific set of base-three tiles is listed for each problem. Your task is to form a rectangle using all of the tiles. Begin at the origin with the flats and build outward, keeping the flats as close to the origin as possible. After you have used all the flats, use the longs and then the units.

a. Sketch a picture of each resulting rectangle, showing the four sections occupied by flats, longs, and units.
b. From the layout, determine the length and width in numerical and algebraic forms using expanded notation.
c. Apply the distributive property in multiplying length x width to determine the area. Collect the terms.
d. Compare the result of your computation with the picture.

1. 1 flat, 2 longs, 1 unit

a. Sketch

<u>length</u>   <u>width</u>   <u>area</u>

b. (   )(   ) = (3^2) +   (3) +   (3) +

=

c. (   )(   ) = x^2 + x + x +

=

2. 1 flat, 3 longs, 2 units

a. Sketch

<u>length</u>   <u>width</u>   <u>area</u>

b. (   )(   ) = (3^2) +   (3) +   (3) +

=

c. (   )(   ) = x^2 + x + x +

=

3. 2 flats, 5 longs, 2 units

a. Sketch

<u>length</u>   <u>width</u>   <u>area</u>

b. (   )(   ) = (3^2) +   (3) +   (3) +

=

c. (   )(   ) = x^2 + x + x +

=

4. 2 flats, 6 longs, 4 units

a. Sketch

<u>length</u>   <u>width</u>   <u>area</u>

b. (   )(   ) = (3^2) +   (3) +   (3) +

=

c. (   )(   ) = x^2 + x + x +

=

5. 3 flats, 5 longs, 2 units
a. Sketch

length   width   area

b. $($ $\quad$ $)($ $\quad$ $)=$ $(3^2)\ +$ $\ (3)\ +$ $\ (3)\ +$

$\quad\quad\quad\quad\quad\quad\quad\quad\quad =$

c. $($ $\quad$ $)($ $\quad$ $)=$ $x^2\ +$ $\ x\ +$ $\ x\ +$

$\quad\quad\quad\quad\quad\quad\quad\quad\quad =$

6. 3 flats, 7 longs, 2 units
a. Sketch

length   width   area

b. $($ $\quad$ $)($ $\quad$ $)=$ $(3^2)\ +$ $\ (3)\ +$ $\ (3)\ +$

$\quad\quad\quad\quad\quad\quad\quad\quad\quad =$

c. $($ $\quad$ $)($ $\quad$ $)=$ $x^2\ +$ $\ x\ +$ $\ x\ +$

$\quad\quad\quad\quad\quad\quad\quad\quad\quad =$

7. 3 flats, 8 longs, 4 units
a. Sketch

length   width   area

b. $($ $\quad$ $)($ $\quad$ $)=$ $(3^2)\ +$ $\ (3)\ +$ $\ (3)\ +$

$\quad\quad\quad\quad\quad\quad\quad\quad\quad =$

c. $($ $\quad$ $)($ $\quad$ $)=$ $x^2\ +$ $\ x\ +$ $\ x\ +$

$\quad\quad\quad\quad\quad\quad\quad\quad\quad =$

8. 4 flats, 4 longs, 1 unit
a. Sketch

length   width   area

b. $($ $\quad$ $)($ $\quad$ $)=$ $(3^2)\ +$ $\ (3)\ +$ $\ (3)\ +$

$\quad\quad\quad\quad\quad\quad\quad\quad\quad =$

c. $($ $\quad$ $)($ $\quad$ $)=$ $x^2\ +$ $\ x\ +$ $\ x\ +$

$\quad\quad\quad\quad\quad\quad\quad\quad\quad =$

9. 4 flats, 6 longs, 2 units
a. Sketch

length   width   area

b. $($ $\quad$ $)($ $\quad$ $)=$ $(3^2)\ +$ $\ (3)\ +$ $\ (3)\ +$

$\quad\quad\quad\quad\quad\quad\quad\quad\quad =$

c. $($ $\quad$ $)($ $\quad$ $)=$ $x^2\ +$ $\ x\ +$ $\ x\ +$

$\quad\quad\quad\quad\quad\quad\quad\quad\quad =$

PAR 4 Constructions

A specific set of base four tiles is listed for each problem. Your task is to form a rectangle using all of the tiles. Begin at the origin with the flats and build outward, keeping the flats as close to the origin as possible. After you have used all the flats, use the longs and finish by using the units.

a. Sketch a picture of each resulting rectangle, showing the four sections occupied by flats, longs, and units.
b. From the layout of the tiles, determine the length and width in both numerical and algebraic form using expanded notation.
c. Apply the distributive property in multiplying length x width to determine the area. Collect the terms.
d. Compare the result of your computation with the picture.

1. 1 flat, 3 longs, 2 units
a. Sketch

length   width   area

b. (   )(   ) = (4^2) + (4) + (4) +
                        =
c. (   )(   ) = x^2 + x + x +
                        =

2. 1 flat, 6 longs, 9 units
a. Sketch

length   width   area

b. (   )(   ) = (4^2) + (4) + (4) +
                        =
c. (   )(   ) = x^2 + x + x +
                        =

3. 2 flats, 7 longs, 6 units
a. Sketch

length   width   area

b. (   )(   ) = (4^2) + (4) + (4) +
                        =
c. (   )(   ) = x^2 + x + x +
                        =

4. 2 flats, 8 longs, 6 units
a. Sketch

length   width   area

b. (   )(   ) = (4^2) + (4) + (4) +
                        =
c. (   )(   ) = x^2 + x + x +
                        =

MULTIPLICATION THE ALGEBRA WAY   152

5. 3 flats, 9 longs, 6 units

a. Sketch

length	width	area

b. (　　　)(　　　) = (4²) + (4) + (4) +
　　　　　　　　　=

c. (　　　)(　　　) = x^2 + x + x +
　　　　　　　　　=

---

6. 2 flats, 7 longs, 3 units

a. Sketch

length	width	area

b. (　　　)(　　　) = (4²) + (4) + (4) +
　　　　　　　　　=

c. (　　　)(　　　) = x^2 + x + x +
　　　　　　　　　=

---

7. 3 flats, 5 longs, 2 units

a. Sketch

length	width	area

b. (　　　)(　　　) = (4²) + (4) + (4) +
　　　　　　　　　=

c. (　　　)(　　　) = x^2 + x + x +
　　　　　　　　　=

---

8. 3 flats, 7 longs, 2 unit

a. Sketch

length	width	area

b. (　　　)(　　　) = (4²) + (4) + (4) +
　　　　　　　　　=

c. (　　　)(　　　) = x^2 + x + x +
　　　　　　　　　=

---

9. 4 flats, 10 longs, 4 units

a. Sketch

length	width	area

b. (　　　)(　　　) = (4²) + (4) + (4) +
　　　　　　　　　=

c. (　　　)(　　　) = x^2 + x + x +
　　　　　　　　　=

A specific set of base five tiles is listed for each problem. Your task is to form a rectangle using all of the tiles. Begin at the origin with the flats and build outward, keeping the flats as close to the origin as possible. After you have used all the flats, use the longs and finish by using the units.

a. Sketch a picture of each resulting rectangle, showing the four sections occupied by flats, longs, and units.
b. From the layout of the tiles, determine the length and width in both numerical and algebraic form using expanded notation.
c. Apply the distributive property in multiplying length x width to determine the area. Collect the terms.
d. Compare the result of your computation with the picture.

1. 1 flat, 6 longs, 9 units
a. Sketch

length	width	area

b. $($ $)($ $) =$ $(5^2) +$ $(5) +$ $(5) +$
$=$

c. $($ $)($ $) =$ $x^2 +$ $x +$ $x +$
$=$

2. 2 flat, 7 longs, 6 units
a. Sketch

length	width	area

b. $($ $)($ $) =$ $(5^2) +$ $(5) +$ $(5) +$
$=$

c. $($ $)($ $) =$ $x^2 +$ $x +$ $x +$
$=$

3. 2 flats, 7 longs, 3 units
a. Sketch

length	width	area

b. $($ $)($ $) =$ $(5^2) +$ $(5) +$ $(5) +$
$=$

c. $($ $)($ $) =$ $x^2 +$ $x +$ $x +$
$=$

4. 2 flats, 4 longs, 2 units
a. Sketch

length	width	area

b. $($ $)($ $) =$ $(5^2) +$ $(5) +$ $(5) +$
$=$

c. $($ $)($ $) =$ $x^2 +$ $x +$ $x +$
$=$

5. 3 flats, 9 longs, 6 units
a. Sketch

length	width		area		
b. ( ) ( ) =			(5^2) +	(5) +	(5) +
=					
c. ( ) ( ) =			x^2 +	x +	x +
=					

6. 3 flats, 6 longs, 3 units
a. Sketch

length	width		area		
b. ( ) ( ) =			(5^2) +	(5) +	(5) +
=					
c. ( ) ( ) =			x^2 +	x +	x +
=					

7. 3 flats, 5 longs, 2 units
a. Sketch

length	width		area		
b. ( ) ( ) =			(5^2) +	(5) +	(5) +
=					
c. ( ) ( ) =			x^2 +	x +	x +
=					

8. 3 flats, 8 longs, 4 unit
a. Sketch

length	width		area		
b. ( ) ( ) =			(5^2) +	(5) +	(5) +
=					
c. ( ) ( ) =			x^2 +	x +	x +
=					

9. 3 flats, 7 longs, 2 units
a. Sketch

length	width		area		
b. ( ) ( ) =			(5^2) +	(5) +	(5) +
=					
c. ( ) ( ) =			x^2 +	x +	x +
=					

Connecting Learning

1. How was this activity like others that you have done? How was it different?

2. How did the numeric and algebraic forms of the equations compare?

3. How did the results of your computations compare with the sketches you made of each set of tiles?

4. What changed about your equations when there was more than one flat?

5. What did you learn about the distributive property from this activity?

MYSTERY NUMBERS

Topic
Distributive property

Key Question
How can the value of *x* be determined given the area of a rectangle and its layout of flats, longs, and units?

Learning Goals
Students will:
- determine the length, width, and area of a rectangle in terms of *x* from a picture,
- write an algebraic equation in terms of *x* for length x width = area and collect terms,
- use trial and error to find the value of *x*,
- translate the equation into numerical values,
- use the distributive property to find the four numerical subproducts, and
- compare the numerical and algebraic mathematical sentences.

Guiding Documents
Project 2061 Benchmark
- *Mathematical ideas can be represented concretely, graphically, and symbolically.*

Common Core State Standards for Mathematics*
- *Make sense of problems and persevere in solving them. (MP.1)*
- *Construct viable arguments and critique the reasoning of others. (MP.3)*
- *Use appropriate tools strategically. (MP.5)*
- *Look for and make use of structure. (MP.7)*
- *Work with radicals and integer exponents. (8.EE.A)*

Math
Distributive property
Multiplication
Expanded notation
Base *x* and mystery bases

Integrated Processes
Observing
Collecting and recording data
Generalizing
Comparing

Materials
Student sheets

Background Information
In this activity, students are given two items of information: the tile layout of a rectangle and the numerical value of the area. From this they are to determine the base of the blocks used in constructing the rectangle.

This activity builds several bridges between the numbers of arithmetic and the literal notation of algebra in the context of the distributive property. Repeatedly walking over these bridges should help them form a clear understanding of the parallel nature of arithmetic and algebraic expressions and algebra. The fact that *x* can represent different numbers helps lead students to an understanding of algebra as generalized arithmetic.

At this stage, they will use trial and error to solve for the mystery base. This will involve substituting different values for *x* in the equation derived from the picture to determine which completes a true statement. In the process, students will have extensive practice in translating from an unknown *x* to a numerical value. Once the numerical value of *x* is determined, they will write a parallel equation in terms of that value, apply the distributive property, and test whether the substitution gives the stated area.

Management
1. Students will need to be familiar with number bases and the distributive property before beginning this activity.
2. There are two parts to this activity that are essentially the same. The second section has been provided for additional practice, and is not necessary if students have a good grasp on the concepts after completing part one.

Procedure
1. Hand out the student sheets for the first section and go over the instructions. Be sure that students understand what to do in each section, including the fact that they will be using trial and error to determine the base of the equations.
2. Have students work together in groups to complete the equations. If desired, give them *Mystery Numbers II* when they are finished.
3. After all groups have completed the student sheets, close with a time of class discussion and sharing.

Connecting Learning

1. How is this activity similar to others you have done? How is it different?
2. Was it difficult to determine which base the equation was in? Explain.
3. Did the sum from the numerical components match the sum from the algebraic components? [Yes.] Why? [The sum is how you determine the base. If the base was determined correctly, the sums will always be the same.]
4. What did this activity teach you about number bases?

Extensions

1. The same students sheets can be used to give rise to additional problems. Simply use values of x other than those in each activity and compute the resulting area of the rectangle. In this way the student pages can be used over and over again without repeating any problems.
2. Challenge students to create their own problems to trade with classmates to solve.

Solutions

The solutions for both parts of *Mystery Numbers* are shown below.

Mystery Numbers 1

1. $(x + 4)(x + 3)$
   $= x^2 + 3x + 4x + 12$
   $= x^2 + 7x + 12 = 90$
   $x = 6$
   $(6 + 4)(6 + 3)$
   $= 6^2 + 3(6) + 4(6) + 12 = 90$

2. $(2x + 2)(x + 2)$
   $= 2x^2 + 4x + 2x + 4$
   $= 2x^2 + 6x + 4 = 144$
   $x = 7$
   $(2 \cdot 7 + 2)(7 + 2)$
   $= 2(7^2) + 4(7) + 2(7) + 4 = 144$

3. $(x + 5)(x + 3)$
   $= x^2 + 3x + 5x + 15$
   $= x^2 + 8x + 15 = 168$
   $x = 9$
   $(9 + 5)(9 + 3)$
   $= 9^2 + 3(9) + 5(9) + 15 = 168$

4. $(2x + 2)(x + 3)$
   $= 2x^2 + 6x + 2x + 6$
   $= 2x^2 + 8x + 6 = 198$
   $x = 8$
   $(2 \cdot 8 + 2)(8 + 3)$
   $= 2(8^2) + 6(8) + 2(8) + 6 = 198$

5. $(x + 5)(x + 4)$
   $= x^2 + 4x + 5x + 20$
   $= x^2 + 9x + 20 = 132$
   $x = 7$
   $(7 + 5)(7 + 4)$
   $= 7^2 + 4(7) + 5(7) + 20 = 132$

6. $(2x + 5)(x + 2)$
   $= 2x^2 + 4x + 5x + 10$
   $= 2x^2 + 9x + 10 = 136$
   $x = 6$
   $(2 \cdot 6 + 5)(6 + 2)$
   $= 2(6^2) + 4(6) + 5(6) + 10 = 136$

7. $(2x + 4)(x + 5)$
   $= 2x^2 + 10x + 4x + 20$
   $= 2x^2 + 14x + 20 = 260$
   $x = 8$
   $(2 \cdot 8 + 4)(8 + 5)$
   $= 2(8^2) + 10(8) + 4(8) + 20 = 260$

8. $(2x + 3)(x + 3)$
   $= 2x^2 + 6x + 3x + 9$
   $= 2x^2 + 9x + 9 = 252$
   $x = 9$
   $(2 \cdot 9 + 3)(9 + 3)$
   $= 2(9^2) + 6(9) + 3(9) + 9 = 252$

9. $(2x + 4)(x + 3)$
   $= 2x^2 + 6x + 4x + 12$
   $= 2x^2 + 10x + 12 = 180$
   $x = 7$
   $(2 \cdot 7 + 4)(7 + 3)$
   $= 2(7^2) + 6(7) + 4(7) + 12 = 180$

Mystery Numbers 2

1. $(x + 5)(x + 4)$
   $= x^2 + 4x + 5x + 20$
   $= x^2 + 9x + 20 = 182$
   $x = 9$
   $(9 + 5)(9 + 4)$
   $= 9^2 + 4(9) + 5(9) + 20 = 182$

2. $(2x + 2)(x + 4)$
   $= 2x^2 + 8x + 2x + 8$
   $= 2x^2 + 10x + 8 = 176$
   $x = 7$
   $(2 \cdot 7 + 2)(7 + 4)$
   $= 2(7^2) + 8(7) + 2(7) + 8 = 176$

3. $(2x + 3)(x + 4)$
   $= 2x^2 + 8x + 3x + 12$
   $= 2x^2 + 11x + 12 = 228$
   $x = 8$
   $(2 \cdot 8 + 3)(8 + 4)$
   $= 2(8^2) + 8(8) + 3(8) + 12 = 228$

4. $(2x + 4)(2x + 2)$
   $= 4x^2 + 4x + 8x + 8$
   $= 4x^2 + 12x + 8 = 224$
   $x = 6$
   $(2 \cdot 6 + 4)(2 \cdot 6 + 2)$
   $= 4(6^2) + 4(6) + 8(6) + 8 = 224$

5. $(3x + 3)(x + 4)$
   $= 3x^2 + 12x + 3x + 12$
   $= 3x^2 + 15x + 12 = 264$
   $x = 7$
   $(3 \cdot 7 + 3)(7 + 4)$
   $= 3(7^2) + 12(7) + 3(7) + 12 = 264$

6. $(3x + 2)(x + 2)$
   $= 3x^2 + 6x + 2x + 4$
   $= 3x^2 + 8x + 4 = 207$
   $x = 7$
   $(3 \cdot 7 + 2)(7 + 2)$
   $= 3(7^2) + 6(7) + 2(7) + 4 = 207$

7. $(2x + 4)(2x + 3)$ $= 4x^2 + 6x + 8x + 12$
                                $= 4x^2 + 14x + 12 = 240$
                     $x = 6$
   $(2 \cdot 6 + 4)(2 \cdot 6 + 3)$ $= 4(6^2) + 6(6) + 8(6) + 12 = 240$

8. $(2x + 4)(3x + 1)$ $= 6x^2 + 2x + 12x + 4$
                                $= 6x^2 + 14x + 4 = 500$
                     $x = 8$
   $(2 \cdot 8 + 4)(3 \cdot 8 + 1)$ $= 6(8^2) + 2(8) + 12(8) + 4 = 500$

9. $(2x + 5)(2x + 3)$ $= 4x^2 + 6x + 10x + 15$
                                $= 4x^2 + 16x + 15 = 483$
                     $x = 9$
   $(2 \cdot 9 + 5)(2 \cdot 9 + 3) = 4(9^2) + 6(9) + 10(9) + 15 = 483$

MYSTERY NUMBERS 1

Pictured below are rectangles constructed with base x blocks. The value of x is the mystery you are to solve.

a. Write an equation in terms of x for length times width = area. Collect like terms.
b. Based on the given value for the area, find the value of x.
c. Translate the length and width into the base that represents the numerical value of x.
d. Use the distributive property to find the four numerical subproducts representing the area.
e. Compare your sum of the numerical components with the given sum for the algebraic components

1. Example

Length	Width		Area

$$(\ x\ +\ 4\ )(\ x\ +\ 3\ )$$

$= x^2 + 3x + 4x + 12$

$= x^2 + 7x + 12 = 90$

$$(\ 6\ +\ 4\ )(\ 6\ +\ 3\ )$$

$x = 6$

$= 6^2 + 3(6) + 4(6) + 12 = 90$

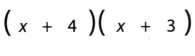

2.

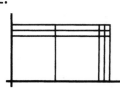

Length	Width		Area

$$(\quad + \quad)(\quad + \quad)$$

$= x^2 + \quad x + \quad x +$

$= x^2 + \quad x + \quad = 144$

$x =$

$$(\quad + \quad)(\quad + \quad)$$

$= \quad + \quad + \quad + \quad =$

3.

Length	Width		Area

$$(\quad + \quad)(\quad + \quad)$$

$= x^2 + \quad x + \quad x +$

$= x^2 + \quad x + \quad = 168$

$x =$

$$(\quad + \quad)(\quad + \quad)$$

$= \quad + \quad + \quad + \quad =$

4.

Area

$= x^2 + \quad x + \quad x +$

Length	Width

$$(\quad + \quad)(\quad + \quad)$$

$= x^2 + \quad x + \quad = 198$

$x =$

$$(\quad + \quad)(\quad + \quad)$$

$= \quad + \quad + \quad + \quad =$

MYSTERY NUMBERS 1

5.

Length	Width		Area

(+   )(+   )   = $x^2 +$ $x +$ $x +$

                                          = $x^2 +$ $x +$ = 132

(+   )(+   ) $x =$

                                          =   +   +   +   =

6.

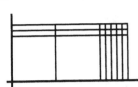

Length	Width		Area

(+   )(+   )   = $x^2 +$ $x +$ $x +$

                                          = $x^2 +$ $x +$ = 136

                       $x =$

(+   )(+   )   =   +   +   +   =

7.

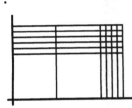

Length	Width		Area

(+   )(+   )   = $x^2 +$ $x +$ $x +$

                                          = $x^2 +$ $x +$ = 260

                       $x =$

(+   )(+   )   =   +   +   +   =

8.

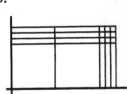

Length	Width		Area

(+   )(+   )   = $x^2 +$ $x +$ $x +$

                                          = $x^2 +$ $x +$ = 252

(+   )(+   )  $x =$

                                          =   +   +   +   =

9.

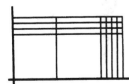

Length	Width		Area

(+   )(+   )   = $x^2 +$ $x +$ $x +$

                                          = $x^2 +$ $x +$ = 180

(+   )(+   ) $x =$

                                          =   +   +   +   =

MYSTERY NUMBERS 2

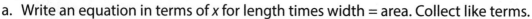

Pictured below are rectangles constructed with base x blocks. The value of x is the mystery you are to solve.

a. Write an equation in terms of x for length times width = area. Collect like terms.
b. Based on the given value for the area, find the value of x.
c. Translate the length and width into the base that represents the numerical value of x.
d. Use the distributive property to find the four numerical subproducts representing the area.
e. Compare your sum of the numerical components with the given sum for the algebraic components.

1.

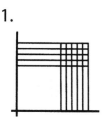

Length	Width		Area
(+)(+)		=	^{2} + + +
		=	^{2} + + = 182
(+)(+)		=	
		=	+ + + =

2.

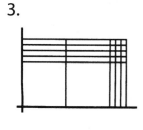

Length	Width		Area
(+)(+)		=	x^2 + x + x +
		=	x^2 + x + = 176
		$x =$	
(+)(+)		=	+ + + =

3.

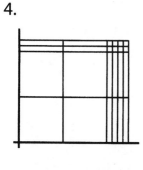

			Area
Length	Width	=	x^2 + x + x +
(+)(+)		=	x^2 + x + = 228
		$x =$	
(+)(+)		=	+ + + =

4.

			Area
		=	x^2 + x + x +
Length	Width	=	x^2 + x + = 224
(+)(+)		$x =$	
		=	+ + + =
(+)(+)			

MYSTERY NUMBERS 2

5.

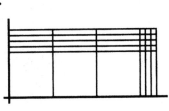

Length	Width		Area
(+)	(+)	=	$x^2 +$ $x +$ $x +$
		=	$x^2 +$ $x +$ = 264
(+)	(+)	$x =$	
		=	+ + + =

6.

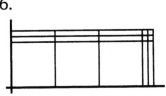

Length	Width		Area
(+)	(+)	=	$x^2 +$ $x +$ $x +$
		=	$x^2 +$ $x +$ = 207
(+)	(+)	$x =$	
		=	+ + + =

7.

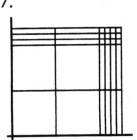

Length	Width		Area
(+)	(+)	=	$x^2 +$ $x +$ $x +$
		=	$x^2 +$ $x +$ = 240
(+)	(+)	$x =$	
		=	+ + + =

8.

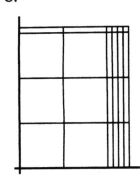

Length	Width		Area
(+)	(+)	=	$x^2 +$ $x +$ $x +$
		=	$x^2 +$ $x +$ = 500
(+)	(+)	$x =$	
		=	+ + + =

9.

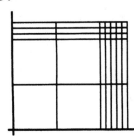

Length	Width		Area
(+)	(+)	=	$x^2 +$ $x +$ $x +$
		=	$x^2 +$ $x +$ = 483
(+)	(+)	$x =$	
		=	+ + + =

MYSTERY NUMBERS

Connecting Learning

1. How is this activity similar to others you have done? How is it different?

2. Was it difficult to determine which base the equation was in? Explain.

3. Did the sum from the numerical components match the sum from the algebraic components? Why?

4. What did this activity teach you about number bases?

Moving Into Algebra

t this point, the transition from arithmetic to algebra becomes complete. The sequence of the first half of this section reflects the development of a multiplication model for the base-ten system used in *Part One*. However, instead of using pieces based on a known length, such as 10, three, four, or five, these pieces are based on an unknown length, *x*.

Filling Quadrilaterals, Building Quadrilaterals

The concept of a variable is introduced in these activities as an unknown length. Beginning in the concrete, students use the pieces to fill or build quadrilaterals.

Modeling Quadrilaterals, Patterns of Special Squares

As students build a number of quadrilaterals, they recognize a pattern of four regions. Dividing a drawing of a quadrilateral into four regions easily represents the four regions. This representation provides an organized method of determining the product of the quadrilateral's dimensions that is reinforced by a memory of meaningful experiences.

The second half of *Part Six* provides a model for factoring to find the dimensions of a quadrilateral of a known area. In this part, the concrete model is always done in the first quadrant. From the concrete experience, students clearly see how the dimensions are found and have strong memories that connect their understanding to the abstract representations.

Parts of Quadrilaterals

This activity provides a transition, starting with the concrete, where students arrange the known pieces into a quadrilateral to discover its dimensions.

Factors from Quadrilaterals, Factor Practice, Factoring Special Squares

As students recognize patterns, they transition to the familiar drawn representation, allowing them to work to an abstract system of solution.

Experience in the classroom has shown that as students move to the representational and abstract levels they have little difficulty dealing with negative terms since they are working at an abstract level. Students will benefit from seeing how negative terms are modeled in four quadrants. (See *Part Seven*.)

The algebra tile pieces used in this section of the book can be provided in one of three formats. The simplest and least expensive is to copy the included template onto card stock and cut the pieces out. Another option is to purchase die cuts for cutting the pieces. Algebra tile materials are also commercially available from AIMS. Whichever format you choose, make sure to have enough materials available for all students.

Filling Quadrilaterals

Topic
Polynomial products

Key Question
How do you determine the dimensions and area of a rectangle by filling it with the fewest algebra tiles?

Learning Goals
Students will:
- fill a rectangle with the fewest number of algebra tiles,
- describe the dimensions of the rectangle as binomials,
- describe the area of a rectangle as a trinomial, and
- recognize that there are always four partial products with the multiplication of two binomials.

Guiding Documents
Project 2061 Benchmark
- *Mathematicians often represent things with abstract ideas, such as numbers or perfectly straight lines, and then work with those ideas alone. The "things" from which they abstract can be ideas themselves (for example, a proposition about "all equal-sided triangles" or "all odd numbers").*

Common Core State Standards for Mathematics*
- *Model with mathematics. (MP.4)*
- *Look for and make use of structure. (MP.7)*
- *Perform arithmetic operations on polynomials. (A-APR.A)*

Math
Algebra
    polynomials
        products
        factors
        distributive property

Integrated Processes
Observing
Comparing and contrasting
Generalizing

Materials
Algebra tiles
Student pages

Background Information
The process of multiplication can be modeled using a rectangle. The dimensions are the factors, and the area is the product. To model the problem 23 times 12, you would make a rectangle 20 + 3 wide and 10 + 2 high. This rectangle would contain four regions. One would be made of two 10 x 10 squares and have an area 200. A second region would be made of three strips. It would measure 10 by 3 and have an area of 30. The third area, made of two strips, would measure 20 by 2 and have an area of 40. The fourth region, made of unit squares, would measure 3 by 2, and have an area of 6. The sum of all four regions would be 276 units.

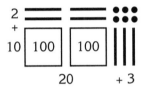

The multiplication of two binomials can be modeled in the same way. Consider the example $(2x + 3)$ by $(x + 2)$. The rectangle's width would be $2x + 3$, and its height would be $x + 2$. The areas of the four regions would be $2x^2$, $3x$, $4x$, and 6, for a total area of $2x^2 + 7x + 6$.

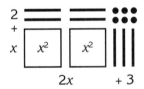

Some similarities are evident between the two examples. Although a different base is used, each of the four partial products is the same. Ten is used as the base in the arithmetic example, and x is used as the base in the algebra example. The resulting products show this similarity—$256 = 2 \cdot 10^2 + 7 \cdot 10 + 6 : 2x^2 + 7x + 6$.

Students can begin to understand the multiplication of binomials by filling in rectangles with algebra tiles. They start by filling the rectangle with as many of the largest pieces (the x-squares) as possible, and then continue using the next smallest pieces (x-strips and unit-squares) until the whole rectangle is covered with the fewest pieces. By filling several rectangles in this way and recording the solutions, students begin to recognize patterns that will help them multiply and factor polynomials. They will have the visual memory that all the rectangles have four regions. There is a region of x-squares. To the right of and above the x-square region there are two x-strip regions. To the right of and above these *x*-strips regions there is a region of unit squares. The areas of these four regions are the partial products of the multiplication of the binomials represented by the dimensions of the rectangle. The sum of the partial products is the product of the two binomials. These patterns will allow the students to move to a representational level in multiplication of binomials.

Management
1. Choose a material to use for algebra tiles and prepare enough for class use.
2. Student teams of two to four work well for these activities. If using the template, have each student cut out a sheet of materials so the group will have enough if they share.
3. To get consistent fill patterns, have all the students begin covering each rectangle from the lower-left hand corner with *x*-squares and work to *x*-strips. Finish covering the rectangles with the unit squares when neither of the other pieces will fit.

Procedure
1. Distribute the algebra tile materials. If students have not used the tiles before, explain the dimensions of each type of piece and its area. The *x*-squares have dimensions of *x* by *x* and have areas of x^2. The *x*-strips are one by *x* and have areas of *x*. The unit squares are one by one and have areas of one.
2. Distribute the rectangles and record sheets. Then discuss with the class strategies for determining the fewest number of pieces used to cover the shaded region.
3. Explain to the class the need to cover the rectangles in a consistent way (see *Management*) so that patterns will become evident. Working together, have the students cover rectangle one. Demonstrate how the rectangle is drawn on the record sheet and have them determine the dimensions and record the area.
4. When students are sure they understand the procedure, have them cover each of the remaining rectangles. When they have completed a solution, have them record a sketch of the arrangement of the pieces and the dimensions and area of the rectangle.

5. When students have filled and recorded all the rectangles, have them discuss similarities and differences and summarize an answer to the *Key Question*.

Connecting Learning
1. How are the dimensions of the rectangles similar and different? [All the dimensions have *x*s plus a number. The number of *x*s and the number added differs in the dimensions.]
2. How are the areas of the rectangles similar and different? [All the areas have x^2s, *x*s, and a number. The number of x^2s, *x*s and the number added differ in the areas.]
3. What are two different ways to describe a rectangle? [dimensions, area]

Extension
Have students construct different rectangles using the materials. Have them trace the outside of the rectangle on one side of a paper. On the other side have the students record a sketch of the pieces making up the rectangle and record the dimensions and area. Students may exchange papers to check each other and get reinforcement.

Solutions

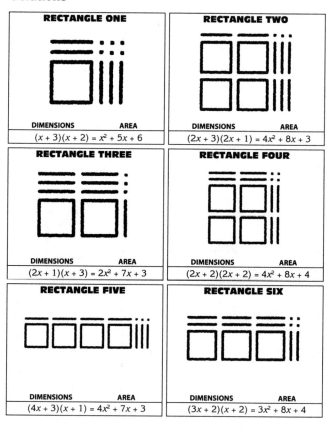

RECTANGLE ONE	
DIMENSIONS	AREA
$(x + 3)(x + 2) = x^2 + 5x + 6$	

RECTANGLE TWO	
DIMENSIONS	AREA
$(2x + 3)(2x + 1) = 4x^2 + 8x + 3$	

RECTANGLE THREE	
DIMENSIONS	AREA
$(2x + 1)(x + 3) = 2x^2 + 7x + 3$	

RECTANGLE FOUR	
DIMENSIONS	AREA
$(2x + 2)(2x + 2) = 4x^2 + 8x + 4$	

RECTANGLE FIVE	
DIMENSIONS	AREA
$(4x + 3)(x + 1) = 4x^2 + 7x + 3$	

RECTANGLE SIX	
DIMENSIONS	AREA
$(3x + 2)(x + 2) = 3x^2 + 8x + 4$	

Filling Quadrilaterals

Algebra Tiles Template

by X X X^2	by X X X^2	by 1 X	by 1 X	by 1 X	by 1 X	by 1 X	by 1 X	by 1 X
by X X X^2	by X X X^2	by 1 X	by 1 X	by 1 X	by 1 X	by 1 X	by 1 X	by 1 X
by X X X^2	by X X X^2	by 1 X	by 1 X	by 1 X	by 1 X	by 1 X	by 1 X	by 1 X
by X 1 X	by X 1 X	1	1	1	1	1	1	1
by X 1 X	by X 1 X	1	1	1	1	1	1	1
by X 1 X	by X 1 X	1	1	1	1	1	1	1
by X 1 X	by X 1 X	1	1	1	1	1	1	1
by X 1 X	by X 1 X	1	1	1	1	1	1	1
by X 1 X	by X 1 X	1	1	1	1	1	1	1

Filling Quadrilaterals

Fill each rectangle with the fewest algebra tiles possible.

1. Start by filling from the lower left corner with the biggest possible pieces (an *x*-square tile).
2. Fill with the *x*-strips next and finally with the unit squares.
3. Make a sketch of how you filled the rectangle on your record sheet.
4. Record the dimensions and area of you record rectangle on your record sheet.

RECTANGLE ONE

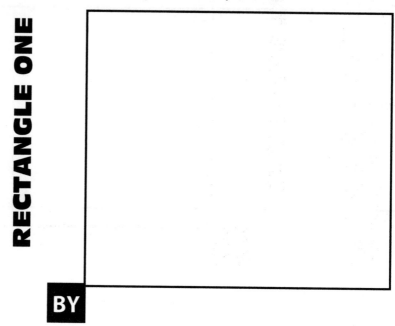

BY

RECTANGLE TWO

BY

Filling Quadrilaterals

RECTANGLE THREE

BY

RECTANGLE FOUR

BY

Filling Quadrilaterals

RECTANGLE FIVE

BY

RECTANGLE SIX

BY

Filling Quadrilaterals

Make a sketch of the way each rectangle was filled and record its dimensions and area.

RECTANGLE ONE

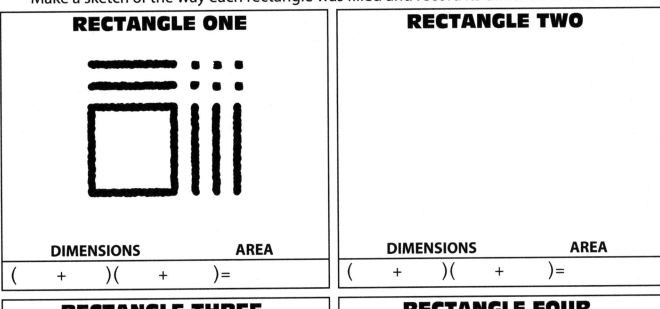

DIMENSIONS   AREA

(+   )(+   )=

RECTANGLE TWO

DIMENSIONS   AREA

(+   )(+   )=

RECTANGLE THREE

DIMENSIONS   AREA

(+   )(+   )=

RECTANGLE FOUR

DIMENSIONS   AREA

(+   )(+   )=

RECTANGLE FIVE

DIMENSIONS   AREA

(+   )(+   )=

RECTANGLE SIX

DIMENSIONS   AREA

(+   )(+   )=

Filling Quadrilaterals

Connecting Learning

1. How are the dimensions of the rectangles similar and different?

2. How are the areas of the rectangles similar and different?

3. What are two different ways to describe a rectangle?

Building Quadrilaterals

Topic
Polynomial products

Key Question
How can you use the dimensions of a rectangle to determine the area of the rectangle?

Learning Goals
Students will:
- learn to describe the dimensions of a rectangle with two binomials,
- learn that the product of two binomials is a trinomial that describes the area of a rectangle, and
- recognize that products of polynomials have four partial products that are represented by the four regions within a rectangle.

Guiding Documents
Project 2061 Benchmark
- *Mathematicians often represent things with abstract ideas, such as numbers or perfectly straight lines, and then work with those ideas alone. The "things" from which they abstract can be ideas themselves (for example, a proposition about "all equal-sided triangles" or "all odd numbers").*

Common Core State Standards for Mathematics*
- *Model with mathematics. (MP.4)*
- *Look for and make use of structure. (MP.7)*
- *Perform arithmetic operations on polynomials. (A-APR.A)*

Math
Algebra
    polynomials
        products
        factors

Integrated Processes
Observing
Comparing and contrasting
Generalizing

Materials
Algebra tiles
Student pages

Background Information
Rectangles can be described in two ways—by dimensions of length and width or by area. Each of the algebra tile pieces can also be described in these two ways. The unit square has the dimensions of one by one and has an area of one square unit. The strip has the dimensions of one by x and has an area of x square units. The x square has the dimensions of x by x and has an area of x^2 square units.

Consider a rectangle with the dimensions of $(3x + 2)$ by $(2x + 2)$ that is built by starting with the largest pieces in the lower left hand corner. The lower left region would be made of six x-squares in a three by two arrangement. The lower right region would by made of four x-strips in a two by two arrangement. The upper left region would be made of six x-strips in a three by two arrangement. The upper right region would be made of four unit squares in a two by two arrangement.

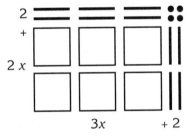

Having students build a number of rectangles with algebra tiles allows them to become familiar with the concepts of factors as the dimensions of a rectangle and the product as the area of a rectangle. As students become familiar with this process, they will recognize the four regions are consistent for all rectangles with binomial dimensions. This understanding will allow student to move from this concrete method of dealing with algebraic multiplication to representational and abstract methods that are based on a meaningful experience.

Management
1. There are two sources of algebra tile materials: student cut materials from the template provided or paper die-cut materials. Secure the desired materials before starting this investigation.
2. So students have easily comparable records, instruct them to build each rectangle using the x-square pieces first starting in the lower left corner. By starting in this corner and building in sequence from largest to smallest, all students will have the same record using the fewest pieces and clearly displaying four regions.

3. Students may find the cross frame useful in building rectangles. The dimensions can be laid out along the bottom and left of the frame using the x-strips and unit squares. The area can then be built in the large upper right quadrant using the dimensions to help verify the size of the rectangle.
4. This activity works best if students are in pairs.
5. A blank student page is included so problems from a student text may be included. The problems must have factors that are binomial and include only addition of positive terms.

Procedure

1. Distribute the student page and discuss the *Key Question* with the class.
2. Provide algebra tiles and cross frames to the students and have them build the rectangles from the dimensions. To ensure that student records are comparable, instruct students to build each rectangle starting in the lower left corner. Have them continue building the rectangle in sequence from the largest to the smallest pieces until the rectangle is completed.
3. Have students record sketches of the pieces in the rectangles and record the areas.
4. Provide time for the students to build all of the rectangles with algebra tile pieces, make sketches, and record the areas.
5. When students have completed all the rectangles, have them compare their solutions to check that they are correct. Then have them discuss what patterns they see in the sketches, dimensions, and areas.

Connecting Learning

1. How are the dimensions of the rectangles similar and different? [All the dimensions have xs plus a number. The number of xs and the number added differ in the dimensions.]
2. How are the areas of the rectangles similar and different? [All the areas have x^2s, xs, and a number. The number of x^2s, xs and the number added differ in the areas.]
3. How are sketches of the rectangles similar and different? [All the sketches have four regions of x^2s, xs, and unit squares. The x^2s are always in the lower left corner of the rectangle. The xs are found both to the right of the x^2s and above the x^2s. The unit squares are found in the upper right corner of the rectangle. Because the areas of the rectangles differ, the number of x^2s, xs, and unit squares differs along with dimensions.]

Extension

Have students arrange the algebra tiles into rectangles of their own design. Then have them record the dimensions onto a sheet paper. On the back of the paper, have the students make a solution key by making a sketch and recording the area of the rectangle. Have the students trade their papers with other students and try to build each other's rectangles.

Solutions

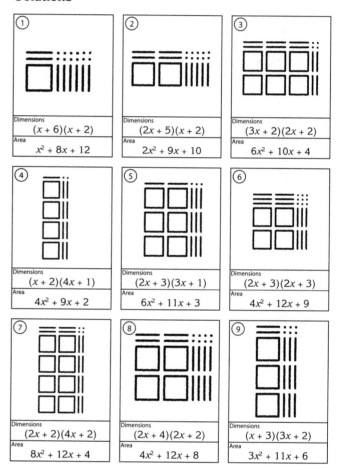

① Dimensions $(x + 6)(x + 2)$ Area $x^2 + 8x + 12$

② Dimensions $(2x + 5)(x + 2)$ Area $2x^2 + 9x + 10$

③ Dimensions $(3x + 2)(2x + 2)$ Area $6x^2 + 10x + 4$

④ Dimensions $(x + 2)(4x + 1)$ Area $4x^2 + 9x + 2$

⑤ Dimensions $(2x + 3)(3x + 1)$ Area $6x^2 + 11x + 3$

⑥ Dimensions $(2x + 3)(2x + 3)$ Area $4x^2 + 12x + 9$

⑦ Dimensions $(2x + 2)(4x + 2)$ Area $8x^2 + 12x + 4$

⑧ Dimensions $(2x + 4)(2x + 2)$ Area $4x^2 + 12x + 8$

⑨ Dimensions $(x + 3)(3x + 2)$ Area $3x^2 + 11x + 6$

Building Quadrilaterals

Cross Frame

X
(by)

Algebra Tiles Template

by X		by X		X by 1	X by 1	X by 1	X by 1	X by 1	X by 1	X by 1
X	X^2	X	X^2	X	X	X	X	X	X	X
by X		by X		X by 1	X by 1	X by 1	X by 1	X by 1	X by 1	X by 1
X	X^2	X	X^2	X	X	X	X	X	X	X
by X		by X		X by 1	X by 1	X by 1	X by 1	X by 1	X by 1	X by 1
X	X^2	X	X^2	X	X	X	X	X	X	X
by 1 X	X	by 1 X	X	1	1	1	1	1	1	1
by 1 X	X	by 1 X	X	1	1	1	1	1	1	1
by 1 X	X	by 1 X	X	1	1	1	1	1	1	1
by 1 X	X	by 1 X	X	1	1	1	1	1	1	1
by 1 X	X	by 1 X	X	1	1	1	1	1	1	1
by 1 X	X	by 1 X	X	1	1	1	1	1	1	1

Building Quadrilaterals

Use the dimensions to build a rectangle. Make a sketch of the rectangle and record the area.

①

Dimensions
$(x + 6)(x + 2)$

Area

②

Dimensions
$(2x + 5)(x + 2)$

Area

③

Dimensions
$(3x + 2)(2x + 2)$

Area

④

Dimensions
$(x + 2)(4x + 1)$

Area

⑤

Dimensions
$(2x + 3)(3x + 1)$

Area

⑥

Dimensions
$(2x + 3)(2x + 3)$

Area

⑦

Dimensions
$(2x + 2)(4x + 2)$

Area

⑧

Dimensions
$(2x + 4)(2x + 2)$

Area

⑨

Dimensions
$(x + 3)(3x + 2)$

Area

MULTIPLICATION THE ALGEBRA WAY

Building Quadrilaterals

Use the dimensions to build a rectangle. Make a sketch of the rectangle and record the area.

①

Dimensions

(+   )(+   )

Area

②

Dimensions

(+   )(+   )

Area

③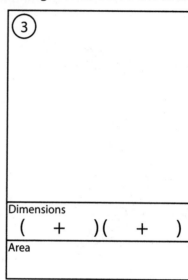

Dimensions

(+   )(+   )

Area

④

Dimensions

(+   )(+   )

Area

⑤

Dimensions

(+   )(+   )

Area

⑥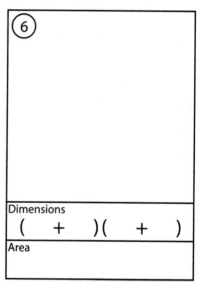

Dimensions

(+   )(+   )

Area

⑦

Dimensions

(+   )(+   )

Area

⑧

Dimensions

(+   )(+   )

Area

⑨

Dimensions

(+   )(+   )

Area

Building Quadrilaterals

Connecting Learning

1. How are the dimensions of the rectangles similar and different?

2. How are the areas of the rectangles similar and different?

3. How are sketches of the rectangles similar and different?

Modeling Quadrilaterals

Topic
Polynomial products

Key Question
How can you determine the area of a rectangle from its dimensions without using algebra tiles?

Learning Goals
Students will:
- represent the four partial products of a rectangle by subdividing a rectangle into four regions,
- learn to record the binomial factors as dimensions along the edges of their rectangle representations, and
- record each partial product in the corresponding region of the rectangle and combine like terms to get the total area of the rectangle.

Guiding Documents
Project 2061 Benchmark
- *Mathematicians often represent things with abstract ideas, such as numbers or perfectly straight lines, and then work with those ideas alone. The "things" from which they abstract can be ideas themselves (for example, a proposition about "all equal-sided triangles" or "all odd numbers").*

Common Core State Standards for Mathematics *
- *Reason abstractly and quantitatively. (MP.2)*
- *Look for and make use of structure. (MP.7)*
- *Perform arithmetic operations on polynomials. (A-APR.A)*

Math
Algebra
    polynomials
        products

Integrated Processes
Observing
Comparing and contrasting
Generalizing

Materials
Student pages

Background Information
By completing the investigations *Filling Quadrilaterals* and *Building Quadrilaterals,* students recognize that binomial dimensions of a rectangle always produce four regions in the rectangle. When students are then asked to visualize such a rectangle, they can be encouraged to draw a rectangle split into four regions. Students will also recall that the lower left region was always made of x-squares, the lower right and upper left regions were made of x-strips, and the upper right region was made of unit squares. With the dimensions written along the edges of the rectangle, the students can record each partial product in the corresponding section of the rectangle. When all the partial products have been recorded, like terms can be combined to get the area as a trinomial.

This representational method of learning to determine the product of two binomials has greater success with students than other abstract methods such as F.O.I.L. for at least three reasons. First, the rectangle representation is based on strong visual and kinesthetic experiences that are easily remembered, so students can reconstruct a meaningful understanding. Second, this representation provides a meaningful representation for factoring trinomials. Third, this representation is easily adapted for determining the product of any polynomial.

All the problems on the student page have positive terms. Students who are familiar with working with positive and negative numbers have little difficulty using them with this representation. Have students do several problems with negative terms out of their textbooks to assess if they transfer their knowledge. If they have difficulty, you might use the investigations *Positives and Negatives* and *Dealing with Negatives* to develop the meaning and effects of negative terms with an area model.

Management
1. Students should have completed *Filling Quadrilaterals* and *Building Quadrilaterals* prior to beginning this investigation.
2. The student page provides generic rectangles for the students to use to practice working out the representations. From the beginning, students can be encouraged to make their own rectangles if only the dimensions are given to them. By the end of the investigation, all students should be able to construct their own rectangle representations.

3. Three student pages are included. The first includes only polynomials with positive terms and addition. The second introduces negative terms and subtraction. A blank student page is included so problems from a student text may be used.

Procedure

1. Have students imagine what a rectangle with the dimensions $(x + 7)$ by $(x + 3)$ looks like. Have students discuss its appearance. Encourage students to recognize that it would have four regions with the x-squares in the lower left region, the x-strips in the lower right and upper left regions, and the unit-squares in the upper right region.
2. Instruct students to make a rectangle and divide it into four regions. Have them record the binomial dimensions along the bottom and left edges of the rectangle, one term for each column and row.
3. Have the students determine and record the area of each region by referring the term at the bottom of the column and the left edge of the row.
4. When students have determined all the partial products, instruct them to combine all the like terms and record the total area of the rectangle.
5. As students become confident in representing the rectangles, have them determine the areas of the rectangles on the student page.

Connecting Learning

1. What do algebraic rectangles have in common? [four regions, x-squares in the lower left, x-strips in the lower right and upper left, unit squares in the upper right]
2. How do you determine the partial product in each region of a rectangle? [Multiply the term at the bottom of the column by the term at the left of the row.]
3. How can you modify the rectangle to multiply a trinomial by a binomial? [Add a third row in the rectangle for the third term in the trinomial. There will be six partial products instead of the four partial products of a binomial multiplied by a binomial.]

Extension

Have students make rectangle representations to complete problems from their textbook that multiply polynomials.

Solutions

1. $(x + 7)(x + 3) = x^2 + 10x + 21$
2. $(2x + 1)(x + 8) = 2x^2 + 17x + 8$
3. $(x + 4)(x + 8) = x^2 + 12x + 32$
4. $(5x + 3)(2x + 1) = 10x^2 + 11x + 3$
5. $(2x + 3)(x + 7) = 2x^2 + 17x + 21$
6. $(13x + 3)(13x + 3) = 169x^2 + 78x + 9$
7. $(3x + 7)(5x + 8) = 15x^2 + 59x + 56$
8. $(4x + 2)(4x + 2) = 16x^2 + 16x + 4$
9. $(x^2 + 5x + 9)(x + 7) = x^3 + 12x^2 + 44x + 63$
10. $(2x^2 + 7x + 11)(3x + 5) = 6x^3 + 31x^2 + 68x + 55$
11. $(x + 7)(x - 3) = x^2 + 4x - 21$
12. $(2x - 1)(x + 8) = 2x^2 + 15x - 8$
13. $(x - 4)(x - 8) = x^2 - 12x + 32$
14. $(5x - 3)(-2x + 1) = -10x^2 + 11x - 3$
15. $(-2x - 3)(x - 7) = -2x^2 + 11x + 21$
16. $(13x - 3)(13x + 3) = 169x^2 - 9$
17. $(3x - 7)(5x + 8) = 15x^2 - 11x - 56$
18. $(4x - 2)(4x - 2) = 16x^2 - 16x + 4$
19. $(x^2 - 5x + 9)(x - 7) = x^3 - 12x^2 + 44x - 63$
20. $(2x^2 - 7x - 11)(3x + 5) = 6x^3 - 11x^2 - 68x - 55$

Modeling Quadrilaterals

Determine the areas of the rectangles. Record the partial products and the polynomial expressions.

Example:

1. $(x + 7)(x + 3) = x^2 + 10x + 21$

+3	3x	21
x	x^2	7x
dimensions	x	+7

2. $(2x + 1)(x + 8) =$ _____

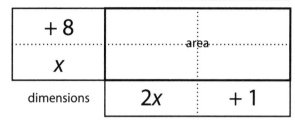

3. $(x + 4)(x + 8) =$ _____

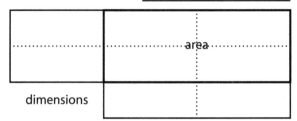

4. $(5x + 3)(2x + 1) =$ _____

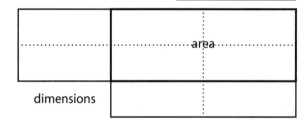

5. $(2x + 3)(x + 7) =$ _____

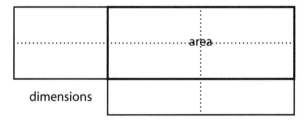

6. $(13x + 3)(13x + 3) =$ _____

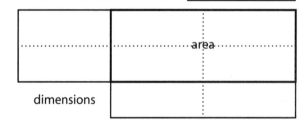

7. $(3x + 7)(5x + 8) =$ _____

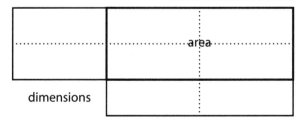

8. $(4x + 2)(4x + 2) =$ _____

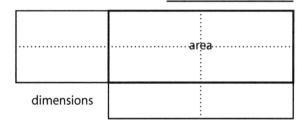

9. $(x^2 + 5x + 9)(x + 7) =$ _____

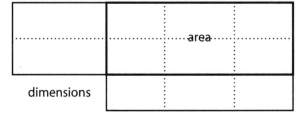

10. $(2x^2 + 7x + 11)(3x + 5) =$ _____

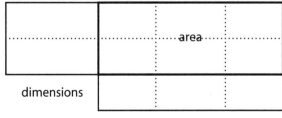

185

Modeling Quadrilaterals

Determine the areas of the rectangles. Record the partial products and the polynomial expressions.

11. $(x + 7)(x - 3) =$ _____

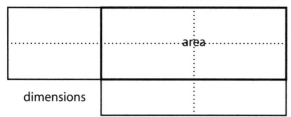

12. $(2x - 1)(x + 8) =$ _____

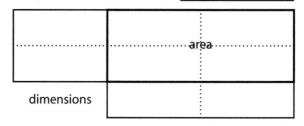

13. $(x - 4)(x - 8) =$ _____

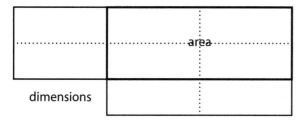

14. $(5x - 3)(-2x + 1) =$ _____

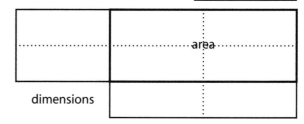

15. $(-2x - 3)(x - 7) =$ _____

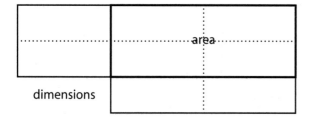

16. $(13x - 3)(13x + 3) =$ _____

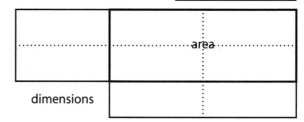

17. $(3x - 7)(5x + 8) =$ _____

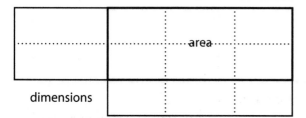

18. $(4x - 2)(4x - 2) =$ _____

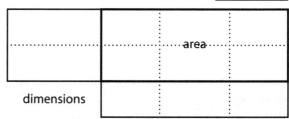

19. $(x^2 - 5x + 9)(x - 7) =$ _____

20. $(2x^2 - 7x - 11)(3x + 5) =$ _____

Modeling Quadrilaterals

Determine the areas of the rectangles. Record the partial products and the polynomial expressions.

1. ( )( ) = _____

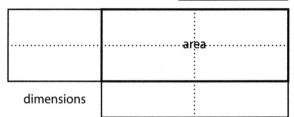

dimensions

2. ( )( ) = _____

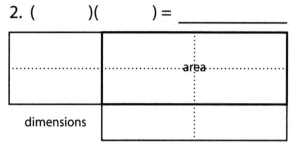

dimensions

3. ( )( ) = _____

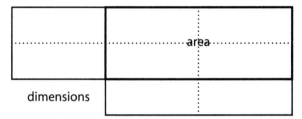

dimensions

4. ( )( ) = _____

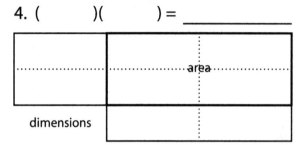

dimensions

5. ( )( ) = _____

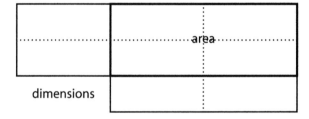

dimensions

6. ( )( ) = _____

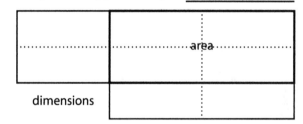

dimensions

7. ( )( ) = _____

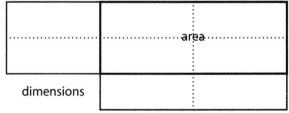

dimensions

8. ( )( ) = _____

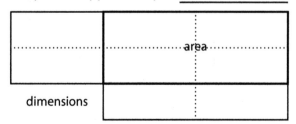

dimensions

9. ( )( ) = _____

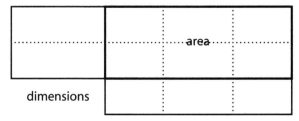

dimensions

10. ( )( ) = _____

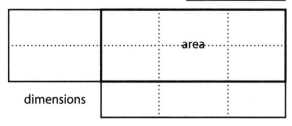

dimensions

Modeling Quadrilaterals

Connecting Learning

1. What do algebraic rectangles have in common?

2. How do you determine the partial product in each region of a rectangle?

3. How can you modify the rectangle to multiply a trinomial by a binomial?

Patterns of Special Squares

Topic
Polynomials

Key Question
What patterns are there in the dimensions and areas of binomial squares?

Learning Goals
Students will:
- determine the product of two binomial squares,
- find the pattern between the dimensions and areas of algebraic squares, and
- learn the algebraic patterns of the special square products.

Guiding Documents
Project 2061 Benchmark
- *Mathematicians often represent things with abstract ideas, such as numbers or perfectly straight lines, and then work with those ideas alone. The "things" from which they abstract can be ideas themselves (for example, a proposition about "all equal-sided triangles" or "all odd numbers").*

Common Core State Standards for Mathematics*
- *Reason abstractly and quantitatively. (MP.2)*
- *Look for and make use of structure. (MP.7)*
- *Perform arithmetic operations on polynomials. (A-APR.A)*

Math
Algebra
    polynomials
        products
            squares of binomials

Integrated Processes
Observing
Comparing and contrasting
Generalizing

Materials
Algebra tiles
Student page

Background Information
Traditionally, when dealing with the products of binomials, students are told the special product patterns and they are expected to memorize them and apply them. With little meaning given to the patterns, the students do not accurately remember them or apply them. This investigation has students practice their skills at finding the products of binomial squares and then seek the patterns that arise. Students who recognize and describe the patterns have a meaningful understanding of where the patterns come from and will have a much longer memory of the algebraic descriptions. Having gone through the work of doing a number of the same type of problems, they will see the purpose of developing a pattern.

The problems on the student page are aligned in three columns of four problems each. The corresponding problems in each column contain the same numbers. The operation between the terms of the binomials in the problems differs by column. An addition operation separates the terms in the first column. A subtraction operation separates the terms in the second column. In the last column, an addition operation separates the terms in one binomial and a subtraction operation separates the terms in other binomial.

As students determine the products, patterns in those products should become evident. All the products in a row contain the same numbers. All the products have a perfect x^2 term and perfect square as the constant term. The middle term is the combination of the two partial products of the two terms in the binomials. The two products are found in the upper left and lower right corners of the representations. In the first column, both products are positive so it is evident that you add the sum of the two products. In the second column, both products are negative so the sums of the two products are subtracted. In the third column, one product is positive and the other negative so they zero each other out leaving the two squares. The resulting patterns are listed below.

Square of a Sum: $(a + b)^2 = a^2 + 2ab + b^2$
Square of a Difference: $(a - b)^2 = a^2 - 2ab + b^2$
Difference of Squares: $(a + b)(a - b) = a^2 - b^2$

The representations on the student page show a generalization of the squares. Since students are working out the products at an abstract level, they refer to the representations to bring meaning to what is happening. Students having difficulty at the abstract level may need to return to the concrete level to make sense of the special squares with subtractions and negative numbers. Consider doing *Dealing with Negatives* and *Positives and Negatives* before doing this investigation.

Management

1. This investigation can easily be completed in one class session. Introducing the investigations at the end of one session can save class time. The calculations can be done as homework and the generalizing of the patterns completed during the second session.
2. Some students may choose to make sketches of the areas of the squares. This visual helps many of the students see the patterns more clearly.
3. If students are having difficulty determine products at the abstract level, have them work through *Dealing with Negatives* and *Positives and Negatives* before doing this investigation.

Procedure

1. Distribute the student page and discuss the *Key Question* with the class.
2. Have students focus on the problems and discuss what similarities and differences they see in the problems in each column and row.
3. Instruct the students to calculate the areas of the squares using the record spaces provided. They may make a sketch if they find it appropriate.
4. When the students have completed all the problems, focus their attention on the products and have them discuss the similarities and differences they see in the products in each column and row.
5. With the type of problem in each column, have students discuss what the patterns are in the products and what causes the patterns.
6. Have students write algebraic descriptions of the patterns. (See *Background Information*.)

Connecting Learning

1. How are the problems in each row similar? [same numbers]
2. How are the problems in each column the same? [same operations]
3. How do the products in each row compare? [same numbers, first and last term perfect squares, third column has only two terms]
4. How do the products in the first column compare? [first and last terms perfect squares, all addition operators, middle term is made of two of the same areas]

5. How do the products in the second column compare? [first and last terms perfect squares, first operator is subtraction and second is addition, middle term is made of two of the same areas]
6. How do the products in the third column compare? [only two terms, both terms perfect squares, all are subtraction]
7. How can you look at the dimensions of a square and know its area? (See *Background Information*.)

Extension

Have students apply their patterns to problems in their algebra books.

Solutions

A1. $(x + 3)^2 = x^2 + 6x + 9$
A2. $(x + 5)^2 = x^2 + 10x + 25$
A3. $(2x + 5)^2 = 4x^2 + 20x + 25$
A4. $(5x + 4)^2 = 25x^2 + 40x + 16$
$(a + b)^2 = a^2 + 2ab + b^2$

B1. $(x - 3)^2 = x^2 - 6x + 9$
B2. $(x - 5)^2 = x^2 - 10x + 25$
B3. $(2x - 5)^2 = 4x^2 - 20x + 25$
B4. $(5x - 4)^2 = 25x^2 - 40x + 16$
$(a - b)^2 = a^2 - 2ab + b^2$

C1. $(x + 3)(x - 3) = x^2 - 9$
C2. $(x + 5)(x - 5) = x^2 - 25$
C3. $(2x + 5)(2x - 5) = 4x^2 - 25$
C4. $(5x + 4)(5x - 4) = 25x^2 - 16$
$(a + b)(a - b) = a^2 - b^2$

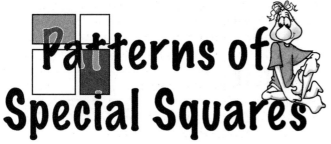

Patterns of Special Squares

Determine the area of the rectangles to see what patterns there are for each type of special square.

A1. $(x + 3)^2 =$ _____

dimensions		area	
x	+3	x	+3

A2. $(x + 5)^2 =$ _____

dimensions | area

A3. $(2x + 5)^2 =$ _____

dimensions | area

A4. $(5x + 4)^2 =$ _____

dimensions | area

Square of a Sum

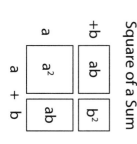 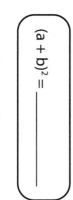 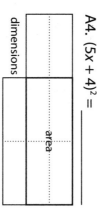

	+b	
a	a^2	ab
+ b	ab	b^2

$(a + b)^2 =$ _____

B1. $(x - 3)^2 =$ _____

dimensions		area	
x	-3	x	- 3

B2. $(x - 5)^2 =$ _____

dimensions | area

B3. $(2x - 5)^2 =$ _____

dimensions | area

B4. $(5x - 4)^2 =$ _____

dimensions | area

Square of a Difference

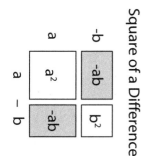  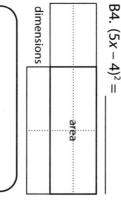

	-b	
a	a^2	-ab
- b	-ab	b^2

$(a - b)^2 =$ _____

C1. $(x + 3)(x - 3) =$ _____

dimensions		area	
x	-3	x	+ 3

C2. $(x + 5)(x - 5) =$ _____

dimensions | area

C3. $(2x + 5)(2x - 5) =$ _____

dimensions | area

C4. $(5x + 4)(5x - 4) =$ _____

dimensions | area

Difference of Squares

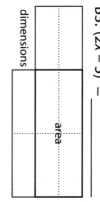 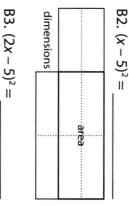

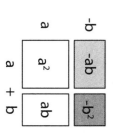  

	-b	
a	a^2	-ab
+ b	ab	$-b^2$

$(a + b)(a - b) =$ _____

Patterns of Special Squares

Connecting Learning

1. How are the problems in each row similar?

2. How are the problems in each column the same?

3. How do the products in each row compare?

4. How do the products in the first column compare?

5. How do the products in the second column compare?

6. How do the products in the third column compare?

7. How can you look at the dimensions of a square and know its area?

PARTS of Quadrilaterals

Topic
Polynomials—factoring

Key Question
How can you determine the dimensions of an algebraic rectangle if all you know is the area?

Learning Goals
Students will:
- gather and arrange a set of algebra tiles into a rectangle given the area,
- identify the dimensions of a algebraic rectangle from the arrangement of algebra tiles, and
- recognize that the dimensions of an algebraic rectangle are the factors of the area.

Guiding Documents
Project 2061 Benchmark
- *Mathematicians often represent things with abstract ideas, such as numbers or perfectly straight lines, and then work with those ideas alone. The "things" from which they abstract can be ideas themselves (for example, a proposition about "all equal-sided triangles" or "all odd numbers").*

Common Core State Standards for Mathematics*
- *Model with mathematics. (MP.4)*
- *Look for and make use of structure. (MP.7)*
- *Perform arithmetic operations on polynomials. (A-APR.A)*

Math
Algebra
    polynomials
        factors

Integrated Processes
Observing
Comparing and contrasting
Generalizing

Materials
Algebra tiles
Student pages

Background Information
Rectangles can be described in two ways—by the dimensions of length and width or by area.

Each of the algebra tile pieces can be described in these two ways. The unit square has the dimensions of one by one and has an area of one square unit. The strip has the dimensions of 1 by x and has an area of x-square units. The x-square has the dimensions of x by x and has an area of x^2 square units.

Consider a situation where there is an x-square, five x-strips, and four unit squares. The only possible rectangular arrangement is to have four x-strips on one side of the x-square and the remaining x-strip above the x-square. The four unit squares would fill the fourth region above the four strips and to the side of the single strip. The dimension along the bottom of the rectangle is $(x + 4)$ and the dimension along the side is $(x + 1)$. This is a model of factoring where given the area or product you determine the dimensions or factors: $x^2 + 5x + 4 = (x + 4)(x + 1)$.

Management
1. Students should already be familiar with finding the area of algebraic rectangles from given dimensions. They are expected to be able to sketch and use the generic rectangular grid representation to make a record and see patterns. They should have completed experiences like *Filling Quadrilaterals, Building Quadrilaterals,* and *Quadrilateral Representations.*
2. Two student pages are included. The first (*Part One*) should be used as the initial experience with factoring. The second page (*Part Two*) should be used after completing *Factors from Quadrilaterals Part One* and *Factor Practice Part One.*

Procedure
1. Distribute the algebra tiles and have students consider the *Key Question* when posed with the first question on the student page.
2. When the students have arranged the four pieces in a square, distribute the student page and have them complete the record.
3. Have the students use the materials to construct each rectangle and then complete the record by making a sketch, filling the rectangular representation, and then completing the equation.
4. When the class has completed the records, have them discuss what patterns they found that may help determine the factors of the area of rectangles.

Connecting Learning

1. How are the rectangles you made similar? [four regions, x-squares, x-strips on side and above x-squares, units diagonal from x-squares]
2. What do you need to consider when placing the x-strips and unit squares? [Consider the ways you can arrange the units into a rectangle and which of those arrangements work with the strips.]
3. Given algebra tiles to work with, how do you determine the dimensions of a rectangle? [Make a rectangle out of all the pieces and the dimensions are the base and height.]

Extension

Have students make their own rectangles out of the tiles. Have them make a record of the area and dimensions along with a sketch on one side of a paper. On the other side, have the students record only the area of the rectangle. Instruct students to exchange papers and looking at the side with only the area, construct the rectangle. To check, students can turn the page over. In this way they can confirm each other's understanding.

Solutions

Part One

1. $(x + 1)(x + 1) = x^2 + 2x + 1$
2. $(x + 1)(x + 2) = x^2 + 3x + 2$
3. $(x + 2)(x + 2) = x^2 + 4x + 4$
4. $(x + 1)(x + 3) = x^2 + 4x + 3$
5. $(x + 2)(x + 3) = x^2 + 5x + 6$
6. $(x + 1)(x + 4) = x^2 + 5x + 4$
7. $(x + 3)(x + 4) = x^2 + 7x + 12$
8. $(x + 2)(x + 4) = x^2 + 6x + 8$
9. $(x + 3)(x + 3) = x^2 + 6x + 9$

Part Two

1. $(x + 1)(2x + 1) = 2x^2 + 3x + 1$
2. $(x + 2)(2x + 1) = 2x^2 + 5x + 2$
3. $(x + 1)(2x + 3) = 2x^2 + 5x + 3$
4. $(x + 2)(2x + 3) = 2x^2 + 7x + 6$
5. $(x + 3)(2x + 1) = 2x^2 + 7x + 3$
6. $(x + 3)(2x + 3) = 2x^2 + 9x + 9$
7. $(2x + 1)(3x + 1) = 6x^2 + 5x + 1$
8. $(2x + 1)(2x + 3) = 4x^2 + 8x + 3$
9. $(2x + 1)(3x + 2) = 6x^2 + 7x + 2$

PARTS of Quadrilaterals
Algebra Tiles Template

by X / X X^2	by X / X X^2	X (by 1, X)	X (by 1, X)	X (by 1, X)	X (by 1, X)	X (by 1, X)	X (by 1, X)	X (by 1, X)
by X / X X^2	by X / X X^2	X (by 1, X)	X (by 1, X)	X (by 1, X)	X (by 1, X)	X (by 1, X)	X (by 1, X)	X (by 1, X)
by X / X X^2	by X / X X^2	X (by 1, X)	X (by 1, X)	X (by 1, X)	X (by 1, X)	X (by 1, X)	X (by 1, X)	X (by 1, X)
by X / 1 X	by X / 1 X	1	1	1	1	1	1	1
by X / 1 X	by X / 1 X	1	1	1	1	1	1	1
by X / 1 X	by X / 1 X	1	1	1	1	1	1	1
by X / 1 X	by X / 1 X	1	1	1	1	1	1	1
by X / 1 X	by X / 1 X	1	1	1	1	1	1	1
by X / 1 X	by X / 1 X	1	1	1	1	1	1	1

1. Parts: 1 x-square
         2 x-strips
         1 square

sketch of quadrilateral

+1	1x	1
x	x^2	1x
dimensions	x	+1

$(x+1)(x+1) = x^2 + 2x + 1$

---

2. Parts: 1 x-square
         3 x-strips
         2 squares

sketch of quadrilateral

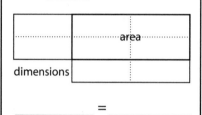

_____ = _____

---

3. Parts: 1 x-square
         4 x-strips
         4 squares

sketch of quadrilateral

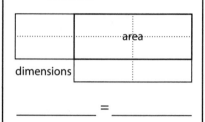

_____ = _____

---

4. Parts: 1 x-square
         4 x-strips
         3 squares

sketch of quadrilateral

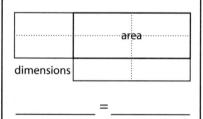

_____ = _____

---

5. Parts: 1 x-square
         5 x-strips
         6 squares

sketch of quadrilateral

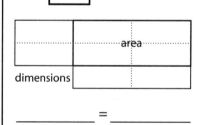

_____ = _____

---

6. Parts: 1 x-square
         5 x-strips
         4 squares

sketch of quadrilateral

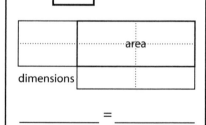

_____ = _____

---

7. Parts: 1 x-square
         7 x-strips
         12 squares

sketch of quadrilateral

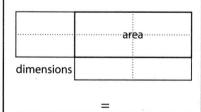

_____ = _____

---

8. Parts: 1 x-square
         6 x-strips
         8 squares

sketch of quadrilateral

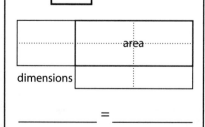

_____ = _____

---

9. Parts: 1 x-square
         6 x-strips
         9 squares

sketch of quadrilateral

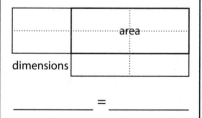

_____ = _____

1. Parts: 2 *x*-squares
            3 *x*-strips
            1 square

sketch of quadrilateral

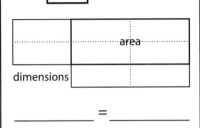

_____ = _____

2. Parts: 2 *x*-squares
            5 *x*-strips
            2 squares

sketch of quadrilateral

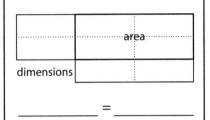

_____ = _____

3. Parts: 2 *x*-squares
            5 *x*-strips
            3 squares

sketch of quadrilateral

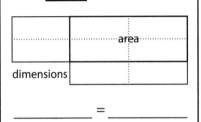

_____ = _____

4. Parts: 2 *x*-squares
            7 *x*-strips
            6 squares

sketch of quadrilateral

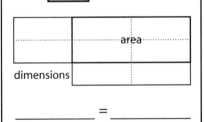

_____ = _____

5. Parts: 2 *x*-squares
            7 *x*-strips
            3 squares

sketch of quadrilateral

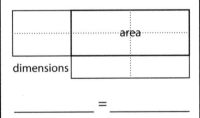

_____ = _____

6. Parts: 2 *x*-squares
            9 *x*-strips
            9 squares

sketch of quadrilateral

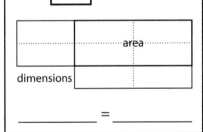

_____ = _____

7. Parts: 6 *x*-squares
            5 *x*-strips
            1 square

sketch of quadrilateral

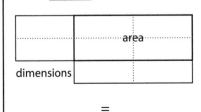

_____ = _____

8. Parts: 4 *x*-squares
            8 *x*-strips
            3 squares

sketch of quadrilateral

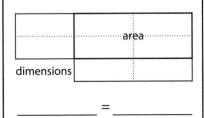

_____ = _____

9. Parts: 6 *x*-squares
            7 *x*-strips
            2 squares

sketch of quadrilateral

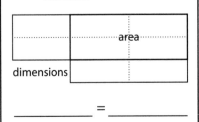

_____ = _____

PARTS of Quadrilaterals

Connecting Learning

1. How are the rectangles you made similar?

2. What do you need to consider when placing the *x*-strips and unit squares?

3. Given algebra tiles to work with, how do you determine the dimensions of a rectangle?

FACTORS FROM QUADRILATERALS

Topic
Polynomials—factors

Key Question
How can you determine the dimensions of an algebraic rectangle if all you know is the area?

Learning Goals
Students will:
- construct, sketch, and symbolically represent rectangles algebraically,
- describe the rectangles with dimensions (factors) and areas (products), and
- find and use patterns from their records to factor polynomials.

Guiding Documents
Project 2061 Benchmark
- *Mathematicians often represent things with abstract ideas, such as numbers or perfectly straight lines, and then work with those ideas alone. The "things" from which they abstract can be ideas themselves (for example, a proposition about "all equal-sided triangles" or "all odd numbers").*

Common Core State Standards for Mathematics*
- *Construct viable arguments and critique the reasoning of others. (MP.3)*
- *Look for and make use of structure. (MP.7)*
- *Perform arithmetic operations on polynomials. (A-APR.A)*

Math
Algebra
    polynomials
        factors

Integrated Processes
Observing
Comparing and contrasting
Generalizing

Materials
Algebra tiles
Student pages

Background Information
Rectangles made with algebra tiles that have areas that are written as trinomials are made up of four regions. There is an x-square region, two x-strip regions, on adjacent sides of the square region, and a unit square region adjacent to both x-strip regions. There

is a relationship between the x-strip regions and the dimensions of the unit square region. This relationship is most evident when only one x-square unit is involved.

Consider all the possible rectangles that have exactly nine x-strips. There are five unique dimension rectangles. The chart emphasizes the patterns that can be found.

Dimensions and area	x-strips in side region	x-strips in top region	unit squares in region
$x(x+9)=x^2+9x$	0	9	0
$(x+1)(x+8)=x^2+9x+8$	1	8	8
$(x+2)(x+7)=x^2+9x+14$	2	7	14
$(x+3)(x+6)=x^2+9x+18$	3	6	18
$(x+4)(x+5)=x^2+9x+20$	4	5	20

In constructing the rectangles, students should recognize that, as the width gets wider by one, the height decreases by one. It also becomes evident that the sum of the two x-strip regions always is nine, the coefficient of the x-term in the area. Also, the product of the two x-strip regions is the number of unit squares because they form the dimensions of the unit square region. These patterns can be used to develop a method for factoring polynomials that is symbolic. Students need to find the two numbers that will give a sum of the coefficient of the x-term and the product of the unit term. These two numbers will be the constants added in the two polynomials of the dimensions.

When more than one x-square is involved, the pattern becomes a bit more complicated, yet it is similar to the preceding pattern. Consider all the rectangles that can be made with three x-squares, 10 x-strips, and any amount of unit squares. The following rectangles are possible.

Dimensions and area	x-strips in side region	x-strips in top region	unit squares in region	x-squares in region
$x(3x+10)=3x^2+10x$	0	10	0	3
$(x+1)(3x+7)=3x^2+10x+7$	3	7	7	3
$(x+2)(3x+4)=3x^2+10x+8$	6	4	8	3
$(x+3)(3x+1)=3x^2+10x+3$	9	1	3	3

It becomes evident that the sum of the two x-strip regions always is 10—the coefficient of the x-term in the area. Also, the product of the two x-strip regions is equal to the product of the number of x-squares and unit squares. Students can use these patterns to find the two numbers that will give a sum of the coefficient of the x-term and the product of the coefficient in the x-square term and the unit term. These two numbers will be number of x-strips in the two regions. To determine the dimensions (factors), the greatest common factor of each row is recorded as a term of one binomial, and the greatest common factor of each column is recorded as a term of the other binomial.

Management

1. Students should already be familiar with finding the area of algebraic rectangles from given dimensions. They are expected to be able to sketch and use the generic rectangular grid representation to make a record and see patterns. They should have completed experiences like *Filling Quadrilaterals, Building Quadrilaterals*, and *Quadrilateral Representations*.

2. Four student pages are included. The first two (*Part One*) should be used as an initial experience with factoring. It should be used after completing *Parts of Quadrilaterals Part One* and before *Factor Practice Part One*. The second set of two pages (*Part Two*) should be used after completing *Parts of Quadrilaterals Part Two* and before *Factor Practice Part Two*.

Procedure

1. Distribute the algebra tiles and have students build all the rectangles that meet the requirements on the student page.

2. Have students record each of the solutions with a sketch, a listing of the dimensions and area of each rectangle, and with a generic rectangle grid.

3. When students have completed both pages, have them look for patterns by answering the questions at the bottom of the second page.

4. When the class has completed the records and looked for patterns, have them discuss what patterns they found and how those patterns can be used to find the dimensions (factors) of a rectangle when they are given the area (product).

Connecting Learning

1. What patterns do you notice in the numbers in the diagonal positions of the charts? (See *Background Information*.)

2. How could you use these patterns to determine the dimensions of a rectangle if you were given the area? (See *Background Information*.)

Solutions

Part One

$x(x + 9) = x^2 + 9x$
$(x + 1)(x + 8) = x^2 + 9x + 8$
$(x + 2)(x + 7) = x^2 + 9x + 14$
$(x + 3)(x + 6) = x^2 + 9x + 18$
$(x + 4)(x + 5) = x^2 + 9x + 20$

Part Two

$(x + 1)(x + 12) = x^2 + 13x + 12$
$(x + 2)(x + 6) = x^2 + 8x + 12$
$(x + 3)(x + 4) = x^2 + 7x + 12$

Part Three

$x(3x + 10) = 3x^2 + 10x$
$(x + 3)(3x + 7) = 3x^2 + 10x + 7$
$(x + 2)(3x + 4) = 3x^2 + 10x + 8$
$(x + 3)(3x + 1) = 3x^2 + 10x + 3$

Part Four

$(x + 3)(2x + 3) = 2x^2 + 9x + 9$
$(x + 9)(2x + 1) = 2x^2 + 19x + 9$
$(x + 1)(2x + 9) = 2x^2 + 11x + 9$

FACTORS FROM QUADRILATERALS

Algebra Tiles Template

by	X	by	X							
X	X^2	X	X^2	X (by 1, X)	X (by 1, X)	X (by 1, X)	X (by 1, X)	X (by 1, X)	X (by 1, X)	X (by 1, X)
by	X	by	X							
X	X^2	X	X^2	X (by 1, X)	X (by 1, X)	X (by 1, X)	X (by 1, X)	X (by 1, X)	X (by 1, X)	X (by 1, X)
by	X	by	X							
X	X^2	X	X^2	X (by 1, X)	X (by 1, X)	X (by 1, X)	X (by 1, X)	X (by 1, X)	X (by 1, X)	X (by 1, X)
by 1, X	X	by 1, X	X	1	1	1	1	1	1	1
by 1, X	X	by 1, X	X	1	1	1	1	1	1	1
by 1, X	X	by 1, X	X	1	1	1	1	1	1	1
by 1, X	X	by 1, X	X	1	1	1	1	1	1	1
by 1, X	X	by 1, X	X	1	1	1	1	1	1	1
by 1, X	X	by 1, X	X	1	1	1	1	1	1	1

FACTORS FROM QUADRILATERALS

PART ONE

Make five rectangles of different dimensions using one x^2 tile, nine x-strips, and any number of unit tiles. Record the sketch of each rectangle. Record the dimensions (factors) and area (product).

Sketch	L x W = Area	Helping Chart
x^2	(+)(+) dimensions / area	dimensions / area
x^2	(+)(+) dimensions / area	dimensions / area
x^2	(+)(+) dimensions / area	dimensions / area
x^2	(+)(+) dimensions / area	dimensions / area
x^2	(+)(+) dimensions / area	dimensions / area

Why can't you make a square out of the parts you had to use?

FACTORS FROM QUADRILATERALS
PART TWO

Make as many different dimension rectangles using one x^2 tile, 12 unit tiles, and any number of x-strips. Record the sketch of each rectangle. Record the dimensions (factors) and area (product).

Sketch	L x W = Area	Helping Chart
x^2	(+)(+) dimensions area	area dimensions
x^2	(+)(+) dimensions area	area dimensions
x^2	(+)(+) dimensions area	area dimensions
x^2	(+)(+) dimensions area	area dimensions

The generalized equation for the rectangles is: $(x + p)(x + q) = x^2 + bx + c$.

1. What pattern do you notice for $(p + q)$ in each equation?

2. What pattern do you notice for $(p \cdot q)$ in each equation?

3. How are these patterns evident in the sketches?

FACTORS FROM QUADRILATERALS
PART THREE

Make as many different dimension rectangles using three *x*-square tiles, 10 *x*-strips, and any number of unit tiles. Record a sketch of each rectangle. Record the dimensions (factors) and area (product).

Sketch	L x W = Area	Helping Chart
	(+)(+) dimensions area	area dimensions
	(+)(+) dimensions area	area dimensions
	(+)(+) dimensions area	area dimensions
	(+)(+) dimensions area	area dimensions
	(+)(+) dimensions area	area dimensions

FACTORS FROM QUADRILATERALS
PART FOUR

Make as many different dimension rectangles using two *x*-square tiles, nine unit tiles, and any number of *x*-strip tiles. Record the sketch of each rectangle. Record the dimensions (factors) and area (product) and complete the chart.

Sketch	**L x W = Area**	**Helping Chart**
	(+)(+) dimensions area	dimensions / area
	(+)(+) dimensions area	dimensions / area
	(+)(+) dimensions area	dimensions / area
	(+)(+) dimensions area	dimensions / area

Consider the area section of each rectangle of the helping chart.

1. What patterns do you notice in the coefficients and constants? (Hint: look at the diagonals of the helping charts)

2. How could you use the patterns to determine the dimensions of a rectangle if you only knew the area?

FACTORS FROM QUADRILATERALS

Connecting Learning

1. What patterns do you notice in the numbers in the diagonal positions of the charts?

2. How could you use these patterns to determine the dimensions of a rectangle if you were given the area?

Factor Practice

Topic
Polynomials—factoring

Key Question
How can you determine the dimensions of an algebraic rectangle without algebra tiles if all you know is the area?

Learning Goals
Students will:
- enter the area of a rectangle (polynomial product) into a generic rectangle,
- use patterns to determine the partial products of the x terms, and
- find the greatest common factors of each pair of partial products to determine the dimensions of the rectangle (factors).

Guiding Documents
Project 2061 Benchmark
- *Mathematicians often represent things with abstract ideas, such as numbers or perfectly straight lines, and then work with those ideas alone. The "things" from which they abstract can be ideas themselves (for example, a proposition about "all equal-sided triangles" or "all odd numbers").*

Common Core State Standards for Mathematics*
- *Reason abstractly and quantitatively. (MP.2)*
- *Look for and make use of structure. (MP.7)*
- *Perform arithmetic operations on polynomials. (A-APR.A)*

Math
Algebra
    polynomials
        factors

Integrated Processes
Observing
Comparing and contrasting
Generalizing

Materials
Student pages

Background Information
Students should understand that algebraic rectangles described by trinomials are made of four regions. All the x-square pieces are in one corner and the unit squares are found in the opposite diagonal corner. The areas of these two regions can easily be recorded in the corresponding corners in a generic rectangle. Consider the trinomial, $6x^2 + 7x + 2$. The $6x^2$ term would be recorded in the lower left corner of the generic rectangle, and two would be recorded in the upper right corner.

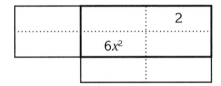

The x-strips divide into two regions, one to the side of the x-squares, and one above the x-squares. The difficulty in factoring is determining how to split the x-strip term into two component parts that will allow factoring. Students who have completed *Factors from Quadrilaterals* should be familiar with the patterns that will allow them to determine how to split the x-strips. The x-strips should be split into two quantities. First, the quantities when multiplied must have the same product as the product of the coefficient of the x-square term and the quantity of unit squares. Second, the two quantities of x-strips must be the total sum of x-strips in the trinomial. For the example, two numbers need to be found that have a product of 12 ($6 \cdot 2$) and a sum of seven. Three and four give a product of 12 and a sum of seven so $3x$ and $4x$ can be recorded in the two remaining regions of the generic rectangle in the upper left and lower right corners.

	$3x$	2
	$6x^2$	$4x$

Now students find the greatest common factor in each row and column of the generic rectangle. In the example, students should find $3x$ and two are the greatest factors in the columns and that $2x$ and one are the greatest factors in the rows. By combining the two terms of greatest common factors, the columns and rows give the dimensions (factors) of the rectangle's area (trinomial product.) $(3x + 2)(2x + 1) = 6x^2 + 7x + 2$

1	3x	2
2x	6x²	4x
	3x	2

All the problems in the *Part One* have all positive components to allow students to become comfortable with the patterns and procedures. Field-testing has shown that once students have become comfortable with the procedure using positive terms, the transition to negative terms causes little difficulty. Some teachers may find it helpful to develop a second page of practice that includes negative terms for their students. *Part Two* introduces negative terms and has students construct their own generic rectangles.

Management

1. Students should already have completed *Factors from Quadrilaterals* to understand the patterns required to factor rectangles.
2. Three student pages are included. *Part One* has only simple trinomials with only one x-square and should be completed after doing *Factors from Quadrilaterals Part One*. *Part Two* includes trinomials with coefficients other than one in the x-square term. *Part Two* should be used after completing *Factors from Quadrilaterals Part Two*. The third page has no problems recorded so that teachers or students can record their own problems.
3. The last problems on the practice pages intentionally do not have generic rectangles to encourage students to make their own records.

Procedure

1. Write the area of the rectangle in the first problem on a board and ask students to consider how the rectangle might look. Elicit the response that there will be four regions with the x-squares being in the lower left, and the unit squares in the upper right corner. Have them record the x-square term and unit term in the appropriate places in the generic rectangle.
2. Encourage the students to recall what pattern there is in how the x-strips should be split to fit into the rectangle. Have students use the pattern to determine how to split the x-strips and have them record the x-strips in the remaining quarters of the generic rectangle.

3. Tell students to determine the greatest common factor of each row and column of the generic rectangle. Have them record the greatest common factors as terms of the binomial factors.
4. Direct students to complete all the problems when they are confident of the pattern and procedure.

Connecting Learning

1. What patterns do all algebraic rectangles share? [four regions, product of x-square coefficient and quantity of unit squares equals the sum of the coefficients of the two x-strip regions]
2. How can you use the patterns of algebraic rectangles to determine the dimensions of the rectangle? [Split the number of x-strips into two groups where the products of the quantities equal the product of the number of x-squares and unit squares.]
3. How do you determine the dimensions of the rectangle if you know the area of each of the four regions? [Find the greatest common factor of each pair of terms (row and column) in the rectangle.]

Extension

Students may use the blank page of generic squares to complete factoring problems found in their textbooks.

Solutions

Part One
1. $(x + 1)(x + 5) = x^2 + 6x + 5$
2. $(x + 4)(x + 4) = x^2 + 8x + 16$
3. $(x + 4)(x + 9) = x^2 + 13x + 36$
4. $(x + 3)(x + 5) = x^2 + 8x + 15$
5. $(x + 1)(x + 4) = x^2 + 5x + 4$
6. $(x + 3)(x + 4) = x^2 + 7x + 12$
7. $(x + 5)(x + 6) = x^2 + 11x + 30$
8. $(x + 3)(x + 8) = x^2 + 11x + 24$
9. $(x + 2)(x + 7) = x^2 + 9x + 14$
10. $(x + 3)(x + 9) = x^2 + 12x + 27$

Part Two
1. $(x + 5)(3x + 1) = 3x^2 + 16x + 5$
2. $(2x + 1)(3x + 1) = 6x^2 + 5x + 1$
3. $(x + 7)(2x + 3) = 2x^2 + 17x + 21$
4. $(2x + 5)(3x - 4) = 6x^2 + 7x - 20$
5. $(x + 2)(5x - 4) = 5x^2 + 6x - 8$
6. $(2x - 1)(4x + 3) = 8x^2 + 2x - 3$
7. $(x - 2)(2x + 1) = 2x^2 - 3x - 2$
8. $(x - 1)(6x + 5) = 6x^2 - x - 5$
9. $(x - 2)(5x + 1) = 5x^2 - 9x - 2$
10. $(x + 10)(2x - 1) = 2x^2 + 19x - 10$

Factor Practice Practice
Part One

Practice
Practice

1 ( )( ) = $x^2 + 6x + 5$

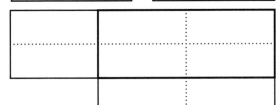

2 ( )( ) = $x^2 + 8x + 16$

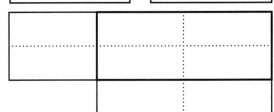

3 ( )( ) = $x^2 + 13x + 36$

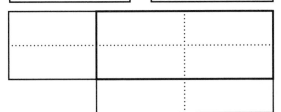

4 ( )( ) = $x^2 + 8x + 15$

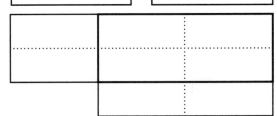

5 ( )( ) = $x^2 + 5x + 4$

6 ( )( ) = $x^2 + 7x + 12$

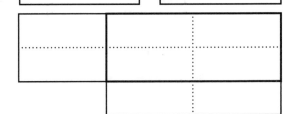

7 ( )( ) = $x^2 + 11x + 30$

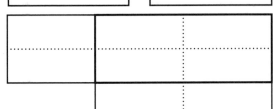

8 ( )( ) = $x^2 + 11x + 24$

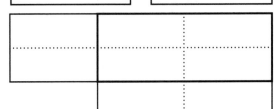

9 ( )( ) = $x^2 + 9x + 14$

10 ( )( ) = $x^2 + 12x + 27$

Factor Practice
Practice **Practice** **Part Two**
Practice

① ( )( ) = $3x^2 + 16x + 5$

② ( )( ) = $6x^2 + 5x + 1$

③ ( )( ) = $2x^2 + 17x + 21$

④ ( )( ) = $6x^2 + 7x - 20$

⑤ ( )( ) = $5x^2 + 6x - 8$

⑥ ( )( ) = $8x^2 + 2x - 3$

⑦ ( )( ) = $2x^2 - 3x - 2$

⑧ ( )( ) = $6x^2 - x - 5$

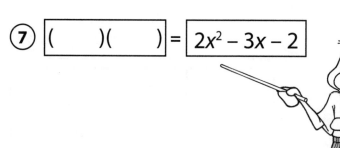

⑨ ( )( ) = $5x^2 - 9x - 2$

⑩ ( )( ) = $2x^2 + 19x - 10$

Factor Practice

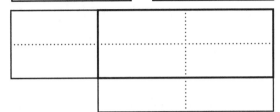

1 ( )( ) =

2 ( )( ) =

3 ( )( ) =

4 ( )( ) =

5 ( )( ) =

6 ( )( ) =

7 ( )( ) =

8 ( )( ) =

9 ( )( ) =

10 ( )( ) =

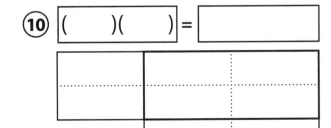

Factor Practice

Connecting Learning

1. What patterns do all algebraic rectangles share?

2. How can you use the patterns of algebraic rectangles to determine the dimensions of the rectangle?

3. How do you determine the dimensions of the rectangle if you know the area of each of the four regions?

FACTORING SPECIAL SQUARES

Topic
Polynomials—factoring

Key Question
How can you recognize algebraic squares to make factoring easier?

Learning Goals
Students will:
- factor a number of algebraic squares, and
- identify patterns that will help them recognize special squares.

Guiding Documents
Project 2061 Benchmark
- *Mathematicians often represent things with abstract ideas, such as numbers or perfectly straight lines, and then work with those ideas alone. The "things" from which they abstract can be ideas themselves (for example, a proposition about "all equal-sided triangles" or "all odd numbers").*

Common Core State Standards for Mathematics*
- *Reason abstractly and quantitatively. (MP.2)*
- *Look for and make use of structure. (MP.7)*
- *Perform arithmetic operations on polynomials. (A-APR.A)*

Math
Algebra
    polynomials
        factors

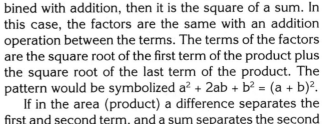

Integrated Processes
Observing
Comparing and contrasting
Generalizing

Materials
Student page

Background Information
Learning to recognize areas that form special squares can make the job of factoring much more efficient. Students who factor a number of areas of the same types of special squares can identify these patterns for themselves.

Areas of perfect squares will have perfect squares for the first and last terms. The middle term will be a product of twice the square root of the first term and the square root of the last term.

If the three terms of the area (product) are combined with addition, then it is the square of a sum. In this case, the factors are the same with an addition operation between the terms. The terms of the factors are the square root of the first term of the product plus the square root of the last term of the product. The pattern would be symbolized $a^2 + 2ab + b^2 = (a + b)^2$.

If in the area (product) a difference separates the first and second term, and a sum separates the second and third terms, the trinomial is a square of a difference. The terms of the factors are the square root of the first term of the product minus the square root of the last term of the product. The pattern would be symbolized $a^2 - 2ab + b^2 = (a - b)^2$.

When the area (product) is described with two terms made of the difference of two squares, the factors are the same except one is a sum and one is a difference. Each factor will have terms of the square root of the first and second terms of the product. One factor will have a sum of the terms, and the other factor will have a difference between the terms. As students factor a number of differences of squares, they will see there is a region of positive x-strips and an equal region of negative x-strips. These two regions zero each other out leaving only the x-square and unit square regions. The pattern of the difference of squares is symbolized $a^2 - b^2 = (a + b)(a - b)$.

Management
1. Students should already completed *Factor Practice* so they can factor the areas on the student page.
2. Class time can be saved by having students do the factoring as homework to be followed up with pattern generalization the next day.

Procedure
1. If necessary, review with students how to find the dimensions of a rectangle when given the area.
2. Distribute the student page and have students compare the problems and identify patterns in the problems.
3. Instruct the students to use the generic rectangles to find the factors for each problem.
4. Have students compare the factors in each column to the corresponding problem to identify patterns.
5. Have students generalize how they can use their patterns to factor similar type problems without generic squares.

Connecting Learning

1. What patterns do you see in the problems in each row? [The first and last terms are the same and are perfect squares. In the first and second columns, the middle term is the same and is twice the square root of the first term and the square root of the last term.]

2. How are all the problems in each column similar? [First column, trinomials with all addition; second column, trinomials with subtraction first, then addition; last column, perfect square terms subtracted from each other.]

3. What patterns do you see in the first column of factors? [Both factors are the same, so the rectangle is a square. The first term of the factor is the square root of the first term of the product. The second term of the factor is the square root of the last term of the product. The terms have an addition sign between them.]

4. What patterns do you see in the second column of factors? [Both factors are the same so the rectangle is a square. The first term of the factor is the square root of the first term of the product. The second term of the factor is the square root of the last term of the product. The terms have a subtraction sign between them.]

5. What patterns do you see in the last column of factors? [Both factors have the same numbers. The first term of the factors is the square root of the first term of the product. The second term of the factor is the square root of the last term of the product. One factor has an addition sign between the terms and the other factor has a subtraction sign.]

Extension

After students have summarized the patterns in algebraic equations, have them apply the patterns to problems found in their textbooks.

Solutions

A1. $(x + 3)^2 = x^2 + 6x + 9$
A2. $(2x + 3)^2 = 4x^2 + 12x + 9$
A3. $(3x + 2)^2 = 9x^2 + 12x + 4$
A4. $(3x + 4)^2 = 9x^2 + 24x + 16$
  $(a + b)^2 = a^2 + 2ab + b^2$

B1. $(x - 3)^2 = x^2 - 6x + 9$
B2. $(2x - 3)^2 = 4x^2 - 12x + 9$
B3. $(3x - 2)^2 = 9x^2 - 12x + 4$
B4. $(3x - 4)^2 = 9x^2 - 24x + 16$
  $(a - b)^2 = a^2 - 2ab + b^2$

C1. $(x + 3)(x - 3) = x^2 - 9$
C2. $(2x + 3)(2x - 3) = 4x^2 - 9$
C3. $(3x + 2)(3x - 2) = 9x^2 - 4$
C4. $(3x + 4)(3x - 4) = 9x^2 - 16$
  $(a + b)(a - b) = a^2 - b^2$

FACTORING SPECIAL SQUARES

Square of a Sum

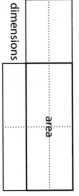

$$(\quad)^2 = a^2 + 2ab + b^2$$

A1. _____ $= x^2 + 6x + 9$

dimensions | area

A2. _____ $= 4x^2 + 12x + 9$

dimensions | area

A3. _____ $= 9x^2 + 12x + 4$

dimensions | area

A4. _____ $= 9x^2 + 24x + 16$

dimensions | area

Square of a Difference

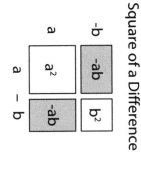

$$(\quad)^2 = a^2 - 2ab + b^2$$

B1. _____ $= x^2 - 6x + 9$

dimensions | area

B2. _____ $= 4x^2 - 12x + 9$

dimensions | area

B3. _____ $= 9x^2 - 12x + 4$

dimensions | area

B4. _____ $= 9x^2 - 24x + 16$

dimensions | area

Difference of Squares

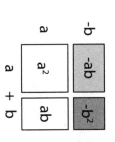

$$(\quad)(\quad) = a^2 - b^2$$

C1. _____ $= x^2 - 9$

dimensions | area

C2. _____ $= 4x^2 - 9$

dimensions | area

C3. _____ $= 9x^2 - 4$

dimensions | area

C4. _____ $= 9x^2 - 16$

dimensions | area

FACTORING SPECIAL SQUARES

Connecting Learning

1. What patterns do you see in the problems in each row?

2. How are all the problems in each column similar?

3. What patterns do you see in the first column of factors?

4. What patterns do you see in the second column of factors?

5. What patterns do you see in the last column of factors?

Dealing With Integers

U p to this point only positive numbers have been modeled and pictured. All representations have taken place in the first quadrant. In the activities that follow, consideration is extended to include negative integers. This requires that physical modeling and pictorial representation must be extended to accommodate negative numbers. In the modeling used by AIMS, this involves using all four quadrants. The article *A New Model for Algebraic Expressions and Equations* on page 5 provides the general background for this activity.

This approach helps students to develop a mental image of what transpires in the use of the distributive property and come to visualize quadratic equations through modeling with algebra tiles.

The activities in *Part Six* assume that students are familiar with the rules of signs for addition and multiplication of integers and have had some experience with collecting like terms. If necessary, these topics should be reviewed before undertaking the activities that follow.

Combinations 1 and 2

In both activities, the problems are presented in pictorial form. Students are asked to construct models of the pictured expressions, physically collect like terms, and make parallel before and after written records. In *Combinations 1,* all terms are positive. Students write expressions in both algebraic and numerical form with x assigned the value of 4. In *Combinations 2,* pictured terms are both positive and negative. Using algebraic notation, students write equations that parallel the conditions before and after collection of terms. Although only the algebraic expressions are requested, students may be instructed to also write the parallel numerical equations.

Positives and Negatives

Using the fewest number of tiles, students fill rectangles on the coordinate plane, identify tiles in each region as positive or negative, pair off and remove congruent positive/negative pairs, and write before and after expressions reflecting the models.

Quadratic Equation Snapshots

Quadratic equations involving both positive and negative integers are pictured. Students translate the pictures into numerical and algebraic expressions and contrast the two forms to see how algebraic expressions relate to arithmetic expressions.

Dealing with Negatives

Students model quadratic equations and the use of the distributive property with both positive and negative integers. They use Quadrants I and II in this activity. Because all three forms are involved, students form associations among the models, their pictorial representations, and the parallel algebraic expressions. In the process, they develop a mental image of progressive steps in the use of the distributive property.

Moving into Four Quadrants

This activity extends *Dealing with Negatives* by using all four quadrants of the coordinate plane.

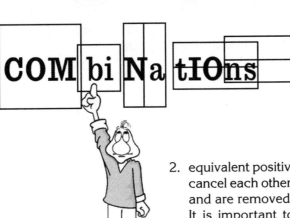

COMbiNatIOns

Topic
Collecting terms

Key Question
How can the process of collecting terms be modeled with base x tiles?

Learning Goals
Students will:
- model the process of collecting positive terms using base x tiles in *Combinations 1*,
- model the process of collecting positive and negative terms using base x tiles in *Combinations 2*, and
- write algebraic equations relating terms before and after they are collected.

Guiding Documents
Project 2061 Benchmark
- *Mathematical ideas can be represented concretely, graphically, and symbolically.*

Common Core State Standards for Mathematics*
- *Model with mathematics. (MP.4)*
- *Use properties of operations to generate equivalent expressions. (7.EE.A)*

Math
Collecting positive and negative terms

Integrated Processes
Observing
Collecting and recording data
Comparing

Materials
Student sheets
Base x tiles (see *Management 1*)

Background Information
In these two activities, students have an opportunity to construct their own understanding of how to collect terms, both positive and negative. They will discover two fundamental ideas that underlie the collection of algebraic terms:
1. like quantities can be combined into a single quantity;

2. equivalent positive and negative quantities in effect cancel each other in the process of being combined and are removed from the resulting expression.

It is important to introduce each activity with a sufficient number of examples so that all students understand the process.

Management
1. Students will need a set of base x tiles in order to model the equations in this activity. Base three, four, five, or ten tiles can be used as any set can take on the identity of base x tiles, as long as there are flats, longs, and units in the set. Each student will need enough tiles to model each of the equations (at least four flats, nine longs, and eleven units).
2. Students should already be familiar with the concepts of collecting terms and negative numbers before beginning this activity.

Procedure
1. Hand out the base x tiles and the student sheet for *Combinations 1* and go over the instructions with the class. *Use your base x tiles to build a model of each picture. Write an expression showing all of the terms to be added, collect like terms, and find the value of the expression if $x = 4$.*
2. Have students work together in groups to complete the first section.
3. When all groups have finished, hand out the student sheets for *Combinations 2* and go over the instructions. *Make a model of each expression, removing the positive and negative pairs that cancel each other out. Draw a picture of the result and write an expression for each side of the equation that identifies positive and negative.*
4. When all groups have finished the second section, close with a time of class discussion and sharing.

Connecting Learning
Combinations 1
1. How did the pictures you created with your tiles compare to the expressions you wrote?
2. How did they compare once the like terms had been collected? [They were the same.]

219

Combinations 2

1. How did the process you went through to combine terms change when there were negative values?
2. Why do positive and negative block pairs cancel each other out?
3. How did using the tiles help you to understand the concept of writing an expression for a picture when there are both positive and negative values?
4. What did you learn about collecting terms from this activity?

Extension

Create a model of the first and second quadrants of the coordinate plane by drawing a line that separates them on a sheet of paper. Have students randomly place flats, longs, and units on each side of the line and then collect the terms. If desired, they can be asked to write an expression for the display before and after collecting the terms as is done in *Combinations 2*.

Solutions

The solutions for both sections are given below. In *Combinations 2* the solutions are given for *Part b* only, with one example at the beginning of how *Part a* should look.

Combinations 1

1. a. $2x + x^2 + 3 + 3x + 2$
    b. $x^2 + 5x + 5 = 41$
2. a. $x^2 + 2 + x^2 + 2x + 4$
    b. $2x^2 + 2x + 6 = 46$
3. a. $3x + 3 + x^2 + 2x + x^2 + 1 + x$
    b. $2x^2 + 6x + 4 = 60$
4. a. $x^2 + 4x + 3 + 2x + x^2 + 3x$
    b. $2x^2 + 9x + 3 = 71$
5. a. $2x^2 + 2x + 2 + x + x^2 + 2x + 4 + x^2$
    b. $4x^2 + 5x + 6 = 90$
6. a. $x + 4x + 2x^2 + 3 + 2x + 2 + x^2$
    b. $3x^2 + 7x + 5 = 81$
7. a. $2x + 2x^2 + 4 + x + x^2 + 3x + 3$
    b. $3x^2 + 6x + 7 = 79$
8. a. $x + 2x^2 + 3x + 2 + 2x^2 + 2x + 4$
    b. $4x^2 + 6x + 6 = 94$

Combinations 2

1. a.

   b. $-x^2 - 2x - 1 + 2x^2 + x + 3 = x^2 - x + 2$
2. $-3x - 2 - x^2 + x + 1 + 2x^2 + 3x = x^2 + x - 1$
3. $-2x^2 - 2x - 3 + x^2 + 4x + 2 = -x^2 + 2x - 1$
4. $-4x - 3 + x^2 + 2x + 5 + x = x^2 - x + 2$
5. $-x^2 - 3x + x^2 + 2x + 2 = -x + 2$
6. $-2x^2 - 2x - x^2 + x^2 + 4x + 1 = -2x^2 + 2x + 1$
7. $-2x - x^2 - 3x - 2 + 2x^2 + 1 + x + 2 + 2x = x^2 - 2x + 1$
8. $-3x - 2x^2 - 2x + x^2 + 2x + 2x^2 = x^2 - 3x$
9. $-3 - x^2 - 4x + 2x + 2 + x^2 + x + 2 = -x + 1$
10. $-2x^2 - 3x - 5 - 2x + 4x + x^2 + 6 = -x^2 - x + 1$
11. $-4x - 2 - 2x - 3 + 2x^2 + x + 2 + 2x + 2 = 2x^2 - 3x - 1$
12. $-x^2 - 3 - 3x - 2 - x + 3x^2 + 1 + 2x + 2 = 2x^2 - 2x - 2$
13. $-3x - 2x^2 - 1 - 2x + x^2 + 3 + 2x + 2 + 2x = -x^2 - x + 4$
14. $-2x^2 - 5 - x^2 - x + x^2 + 3x + 1 + 2x = -2x^2 + 4x - 4$

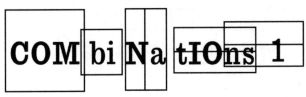

COMbiNatIOns 1

Each picture shows sets composed of the tiles shown at the right. Each term is represented by a single block or a cluster of tiles as shown in the examples.

Use your base x tiles to build a model of each picture, grouping the tiles by kind.

a. Write an expression showing all terms to be added.

b. Collect like terms. Find the value of the expression if $x = 4$.

$1 \quad x \quad x^2$

$2 \quad 3x \quad 2x^2$

1.

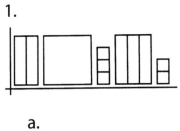

  a.

  b.

2.

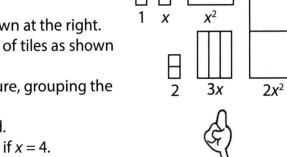

  a.

  b

3.

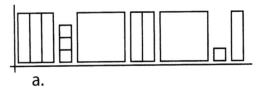

  a.

  b.

4.

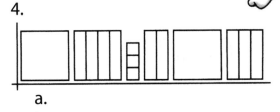

  a.

  b.

5.

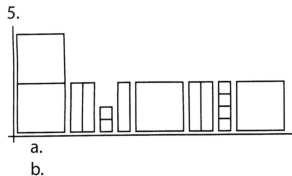

  a.

  b.

6.

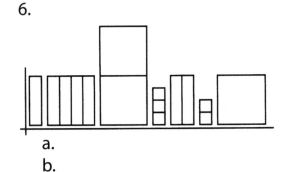

  a.

  b.

7.

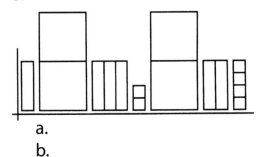

  a.

  b.

8.

  a.

  b.

MULTIPLICATION THE ALGEBRA WAY

© 2012 AIMS Education Foundation

COMbiNatIOns 2

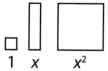
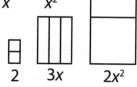

1   x x^2

2   3x $2x^2$

Tiles in the first quadrant represent positive numbers, and those in the second quadrant represent negative numbers. Use your tiles to model each expression pictured below.

a. Positive and negative block pairs cancel each other out. Perform the cancellation by removing such matched pairs. Show the result on the right side of the picture equation.

b. Underneath the picture, write an expression for each side of the equation that identifies the tiles and whether they are positive or negative.

1.

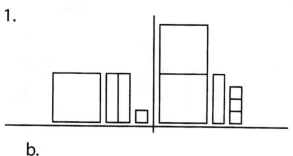

a.

=

b.   =

2.

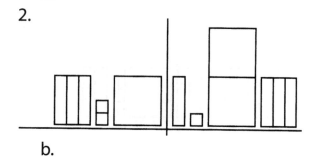

a.

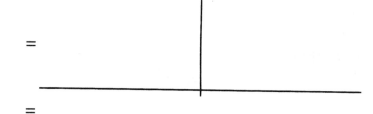

=

b.   =

3.

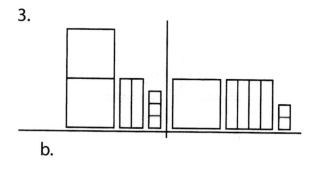

a.

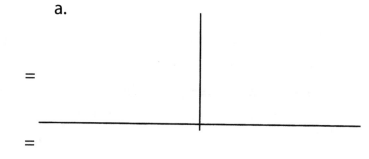

=

b.   =

4.

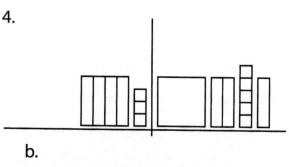

a.

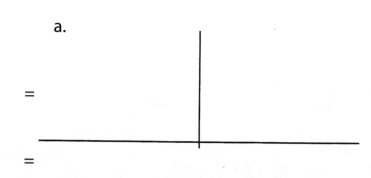

=

b.   =

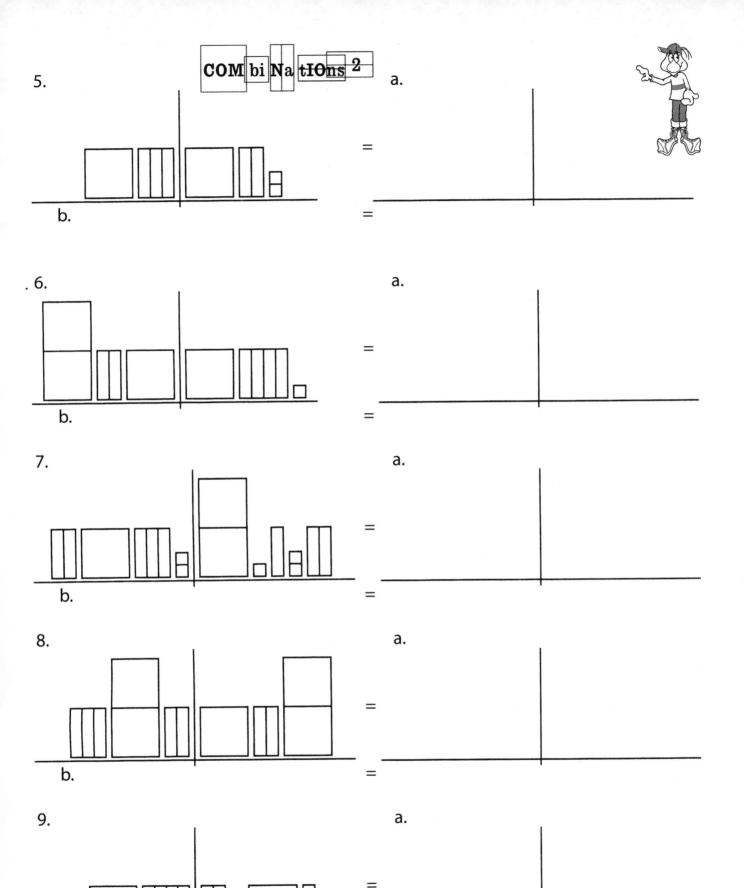

5.

a. _____

= _____

b. _____

= _____

6.

a. _____

= _____

b. _____

= _____

7.

a. _____

= _____

b. _____

= _____

8.

a. _____

= _____

b. _____

= _____

9.

a. _____

= _____

b. _____

= _____

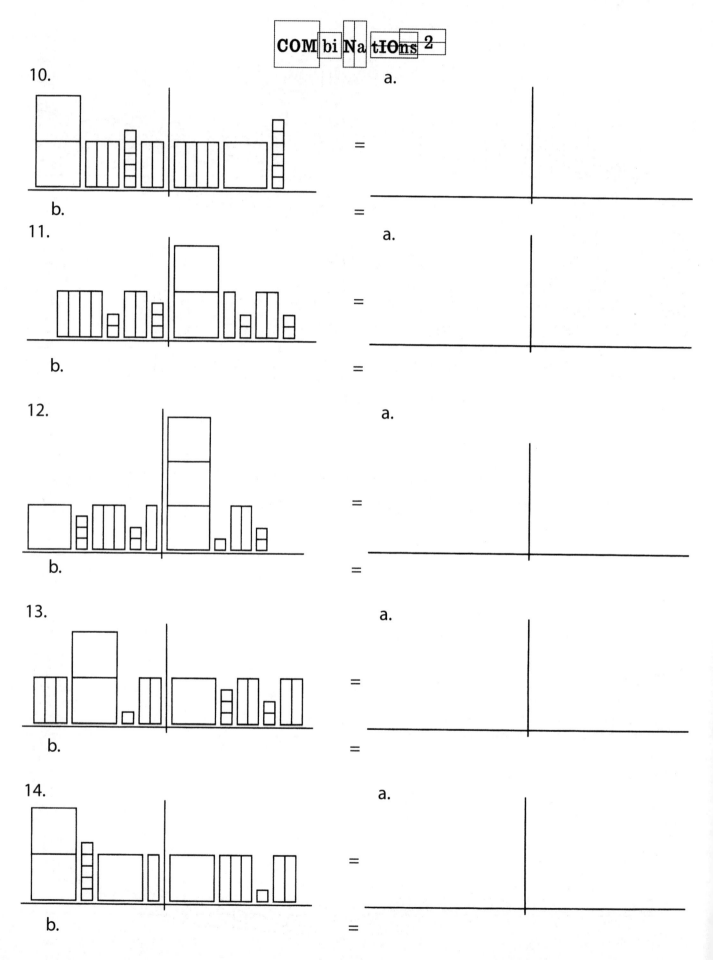

10.

a.

=

b.

=

11.

a.

=

b.

=

12.

a.

=

b.

=

13.

a.

=

b.

=

14.

a.

=

b.

=

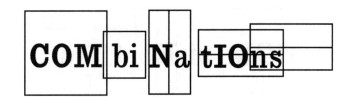

Connecting Learning

Combinations 1

1. How did the pictures you created with your tiles compare to the expressions you wrote?

2. How did they compare once the like terms had been collected?

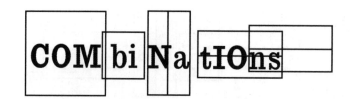

Connecting Learning

Combinations 2

1. How did the process you went through to combine terms change when there were negative values?

2. Why do positive and negative block pairs cancel each other out?

3. How did using the tiles help you to understand the concept of writing an expression for a picture when there are both positive and negative values?

4. What did you learn about collecting terms from this activity?

+ Positives and Negatives –

This activity provides experience with positive and negative quantities using base four blocks to fill in rectangles on the coordinate plane. In the process, students will place blocks in the appropriate positive or negative region, identify blocks in each region as a positive or negative quantity, and pair off and remove congruent positive and negative blocks to model simplification through collection of terms.

Students will identify the dimensions of the rectangle and write sentences showing the dimensions and product of length times width in algebraic terms.

Activities one to six use the first and second quadrant; activities seven and eight take place on the first and fourth quadrant; and activities nine to 12 use all four quadrants. Since such usage will be new to students, it is advisable to model and thoroughly explain several of these problems, or other similar, teacher-designed problems.

*Common Core State Standards for Mathematics**
* *Construct viable arguments and critique the reasoning of others. (MP.3)*
* *Model with mathematics. (MP.4)*
* *Use properties of operations to generate equivalent expressions. (7.EE.A)*

Solutions

1. a.

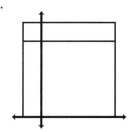

   b. $(x - 1)(x + 1) = x^2 - x + x - 1$
   c. Same as in b with terms in any order.
   d. $(x - 1)(x + 1) = x^2 - 1$

2. a.

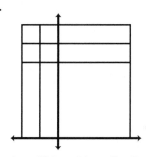

   b. $(x - 2)(x + 2) = x^2 + 2x - 2x - 4$
   c. Same as in b with terms in any order.
   d. $(x - 2)(x + 2) = x^2 - 4$

3. a.

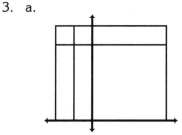

   b. $(x - 2)(x + 1) = x^2 + x - 2x - 2$
   c. Same as in b with terms in any order.
   d. $(x - 2)(x + 1) = x^2 - x - 2$

4. a.

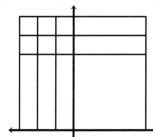

   b. $(x - 3)(x + 2) = x^2 + 2x - 3x - 6$
   c. Same as in b with terms in any order.
   d. $(x - 3)(x + 2) = x^2 - x - 6$

5. a.

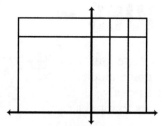

   b. $(-x + 3)(x + 1) = -x^2 - x + 3x + 3$
   c. Same as in b with terms in any order.
   d. $(-x + 3)(x + 1) = -x^2 + 2x + 3$

6. a.

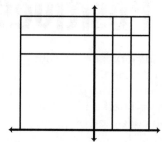

   b. $(-x + 3)(x + 2) = -x^2 - 2x + 3x + 6$
   c. Same as in b with terms in any order.
   d. $(-x + 3)(x + 2) = -x^2 + x + 6$

7. a.

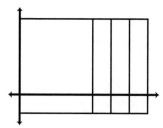

   b. $(x + 3)(x - 1) = x^2 - x + 3x - 3$
   c. Same as in b with terms in any order.
   d. $(x + 3)(x - 1) = x^2 + 2x - 3$

8. a.

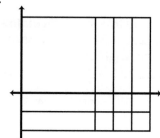

   b. $(x + 3)(x - 2) = x^2 - 2x + 3x - 6$
   c. Same as in b with terms in any order.
   d. $(x + 3)(x - 2) = x^2 + x - 6$

9. a.

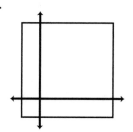

   b. $(x - 1)(x - 1) = x^2 - x - x + 1$
   c. Same as in b with terms in any order.
   d. $(x - 1)(x - 1) = x^2 - 2x + 1$

10. a.

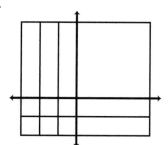

   b. $(x - 3)(x - 2) = x^2 - 2x - 3x + 6$
   c. Same as in b with terms in any order.
   d. $(x - 3)(x - 2) = x^2 - 5x + 6$

11. a.

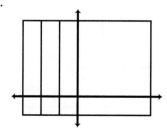

   b. $(x - 3)(x - 1) = x^2 - x - 3x + 3$
   c. Same as in b with terms in any order.
   d. $(x - 3)(x - 1) = x^2 - 4x + 3$

12. a.

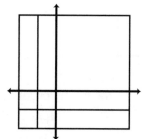

   b. $(x - 2)(x - 2) = x^2 - 2x - 2x + 4$
   c. Same as in b with terms in any order.
   d. $(x - 2)(x - 2) = x^2 - 4x + 4$

+ Positives and Negatives −

a. Fill the outline with the fewest possible blocks so that no block crosses either axis.
b. Make a record of the dimensions of the rectangle and find the dimensions using the distributive property.
c. Write an expression that describes the blocks in the outline. Compare this with your product in (b).
d. Remove all possible matching positive and negative pairs of blocks and write an expression for the result.

1. a.

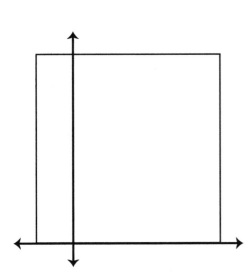

   b. (   )(   ) =

   c.

   d. (   )(   ) =

2. a.

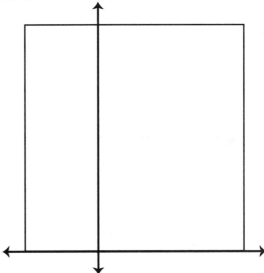

   b. (   )(   ) =

   c.

   d. (   )(   ) =

+ Positives and Negatives –

3. a.

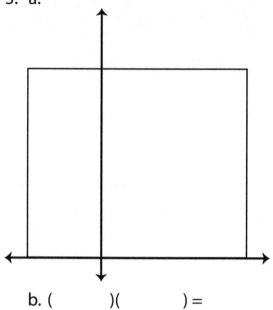

b. (   )(   ) =

c.

d. ((   ) =

4. a.

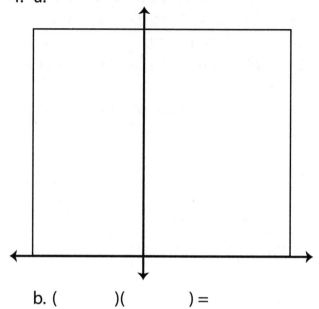

b. (   )(   ) =

c.

d. (   )(   ) =

5. a.

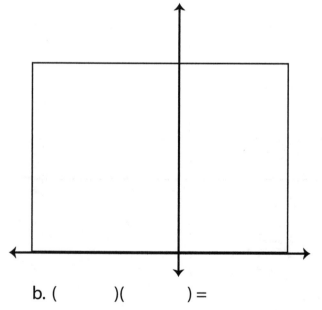

b. (   )(   ) =

c.

d. (   )(   ) =

6. a.

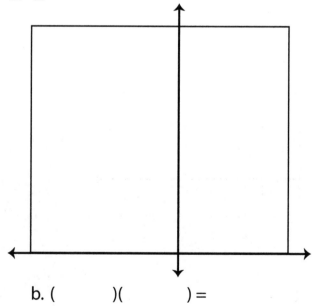

b. (   )(   ) =

c.

d. (   )(   ) =

+ Positives and Negatives –

7. a.

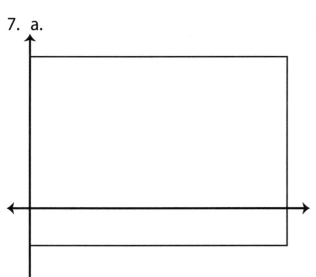

b. (   )(   ) =

c.

d. (   )(   ) =

8. a.

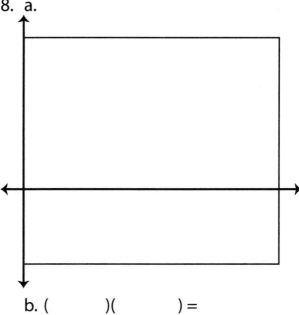

b. (   )(   ) =

c.

d. (   )(   ) =

9. a.

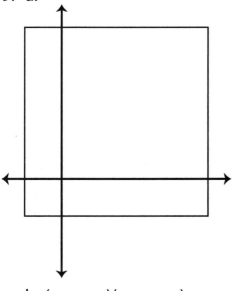

b. (   )(   ) =

c.

d. (   )(   ) =

10. a.

b. (   )(   ) =

c.

d. (   )(   ) =

+ Positives and Negatives −

11.a.

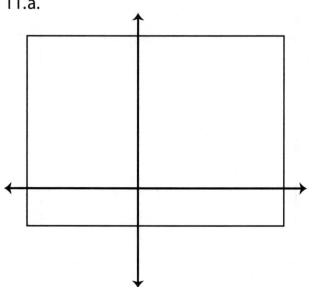

b. ( )( ) =

c.

d. ( )( ) =

12.a.

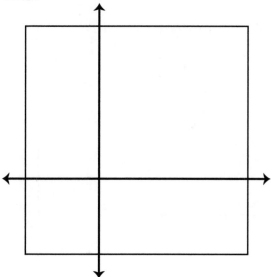

b. ( )( ) =

c.

d. ( )( ) =

Quadratic Equation Snapshots

Topic
Distributive property and quadratic equations

Key Question
How can pictures of quadratic equations be translated into numerical expressions in terms of 5 and algebraic expressions in terms of x?

Learning Goals
Students will:
- translate pictures of quadratic equations into numerical expressions stated in terms of 5,
- translate pictures of quadratic equations into algebraic expressions stated in terms of x, and
- will compare and contrast the two forms to see how algebraic expressions relate to arithmetic expressions.

Guiding Documents
Project 2061 Benchmark
- *Mathematical ideas can be represented concretely, graphically, and symbolically.*

Common Core State Standards for Mathematics*
- *Construct viable arguments and critique the reasoning of others. (MP.3)*
- *Model with mathematics. (MP.4)*
- *Use properties of operations to generate equivalent expressions. (7.EE.A)*

Math
Distributive property
Multiplication
Expanded notation
Collecting terms
Quadratic equations
Bases 5 and x

Integrated Processes
Observing
Recording data
Translating
Comparing
Generalizing

Materials
Student sheets
Base-5 algebra tiles, optional

Background Information
This activity introduces the modeling of quadratic equations involving integers. Both positive and negative quantities must be accommodated. Tiles representing positive quantities are placed in the first and third quadrants and those modeling negative quantities are placed in the second and fourth quadrants.

In this activity, students translate snapshots of quadratic equations into the standard written language of algebra. They are required to associate positive and negative sets of tiles with their symbolic counterparts. This requires careful guidance from the teacher. Understanding the use of such modeling is the foundation for all of the activities involving integers. The advantage of this model is that it mirrors precisely what happens at the abstract level. For example, in adding opposites, the tiles representing equal positive and negative quantities are paired off and removed just as is done at the abstract level.

Students should become familiar with these representations of quadratic equations involving integers by carefully working through as many examples as necessary. Some students will benefit from actually building the model for each rectangle using base-five algebra tiles and relating the manipulation of tiles in each step to what happens at the abstract level.

To break the process into steps, students are asked to identify the set of tiles in each quadrant and the sign that is associated with their placement as shown in the example. The various terms are then collected into an equation. The equation is simplified where possible.

It is important to discuss several algebraic ways of representing the multiplication of two numbers with students, particularly as this occurs in Problems 6, 8, and 10. For example, the product of -2 and 5^2 can be represented as -2×5^2, $-2(5^2)$, or $-2 \cdot 5^2$. The first requires students to know the rules for order of operations when a number is to be added to this product. The second and third require students to know that parenthesis or a dot centered vertically indicate multiplication.

Management
1. You may wish to make base-5 algebra tiles available to students who continue to need the physical model to represent the equations.

Procedure

1. Distribute the student sheets and base-5 algebra tiles, if desired.
2. Have students identify the set(s) of tiles in each quadrant, together with their sign, using literal notation. Instruct them to record this information in the same quadrant as the sets of tiles.
3. Have students identify the length, width, and area of the rectangle in both literal and numeric notation and write the appropriate equations in the spaces provided.
4. Once students have written their equations, they should collect terms wherever possible and compare the numerical and literal statements.

Connecting Learning

1. What term does not always have the same quantity in the algebraic expression as is does in the illustration, the x-square term, the x-term or the unit term? [x-term]
2. In which quadrants were the x-strips when they did not match the quantity in the x-term of the algebraic expression? [quadrant I and quadrant II]
3. Why doesn't the quantity of the x-term in the algebraic expression always match the number of x-strips in the illustration? [each positive x zeros a negative x]

Solutions

The notation for multiplication may vary.

1. $(5 - 3)(5 + 2) = 5^2 + 2(5) - 3(5) - 6 = 14$
   $(x - 3)(x + 2) = x^2 + 2x - 3x - 6 = x^2 - x - 6$

2. $(-5 + 4)(5 + 3) = -5^2 - 3(5) + 4(5) + 12 = -8$
   $(-x + 4)(x + 3) = -x^2 - 3x + 4x + 12 = -x^2 + x + 12$

3. $(5 - 4)(5 + 4)) = 5^2 + 4(5) - 4(5) - 16 = 9$
   $(x - 4)(x + 4) = x^2 + 4x - 4x - 16 = x^2 - 16$

4. $(5 - 4)(5 - 2) = 5^2 - 2(5) - 4(5) + 8 = 3$
   $(x - 4)(x - 2) = x^2 - 2x - 4x + 8 = x^2 - 6x + 8$

5. $(2(5) - 3)(5 - 3) = 2(5^2) - 6(5) - 3(5) + 9 = 14$
   $(2x - 3)(x - 3) = 2x^2 - 6x - 3x + 9 = 2x^2 - 9x + 9$

6. $(-5 + 4)(5 - 1) = -5^2 + 5 + 4(5) - 4 = -4$
   $(-x + 4)(x - 1) = -x^2 + x + 4x - 4 = -x^2 + 5x - 4$

7. $(-2(5) + 3)(5 - 2) = -2(5^2) + 4(5) + 3(5) - 6 = -21$
   $(-2x + 3)(x - 2) = -2x^2 + 4x + 3x - 6 = -2x^2 + 7x - 6$

8. $(-5 + 3)(5 - 3) = -5^2 + 3(5) + 3(5) - 9 = -4$
   $(-x + 3)(x - 3) = -x^2 + 3x + 3x - 9 = -x^2 + 6x - 9$

9. $(2(5) - 4)(5 - 1) = 2(5^2) - 2(5) - 4(5) + 4 = 24$
   $(2x - 4)(x - 1) = 2x^2 - 2x - 4x + 4 = 2x^2 - 6x + 4$

Quadratic Equation Snapshots

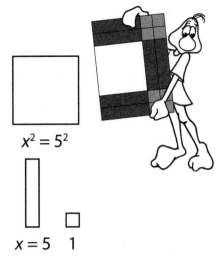

$x^2 = 5^2$

$x = 5$ 1

1. Using algebraic notation, identify the set(s) of tiles in each quadrant together with their sign. Record this information in the same quadrants as the sets of tiles.
2. Identify the length, width, and area and write equations for length x width = area, using both numerical and algebraic notation as shown.
3. Collect terms wherever possible.
4. Compare the numerical and algebraic statements.

Example:

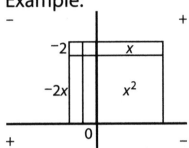

$(5 - 2)(5 + 1) = 5^2 + 5 - 2(5) - 2 = 18$

$(x - 2)(x + 1) = x^2 + x - 2x - 2 = x^2 - x - 2$

1.

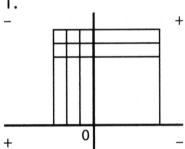

(   )(   ) =

(   )(   ) =

2.

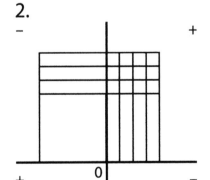

(   )(   ) =

(   )(   ) =

3.

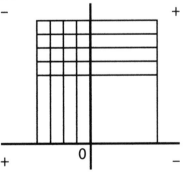

(   )(   ) =

(   )(   ) =

4.

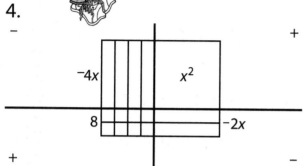

− $-4x$ x^2 +

8 −2x

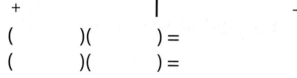

+ −

(   )(   ) =
(   )(   ) =

5.

− +

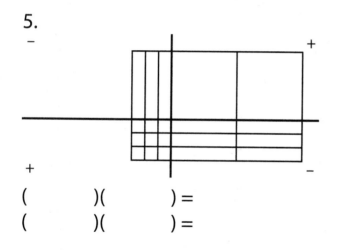

+ −

(   )(   ) =
(   )(   ) =

6.

− +

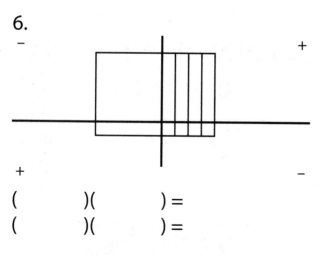

+ −

(   )(   ) =
(   )(   ) =

7.

− +

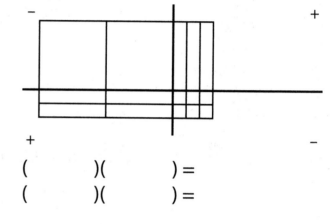

+ −

(   )(   ) =
(   )(   ) =

8.

− +

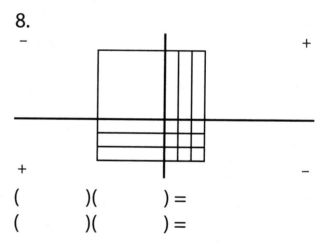

+ −

(   )(   ) =
(   )(   ) =

9.

− +

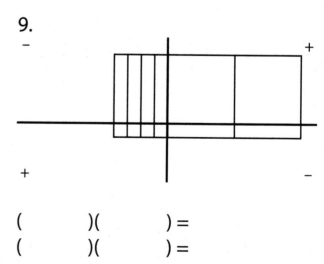

+ −

(   )(   ) =
(   )(   ) =

Connecting Learning

1. What term does not always have the same quantity in the algebraic expression as is does in the illustration, the x-square term, the x-term or the unit term?

2. In which quadrants were the x-strips when they did not match the quantity in the x-term of the algebraic expression?

3. Why doesn't the quantity of the x-term in the algebraic expression always match the number of x-strips in the illustration?

Dealing With Negatives

Topic
Distributive property involving positive and negative integers

Key Question
How can the use of the distributive property involving positive and negative integers be modeled on the coordinate plane?

Learning Goals
Students will:
- become acquainted with modeling the distributive property involving integers;
- build an association between the model, its representation in a sketch, and the parallel algebraic expression; and
- develop a mental image of progressive steps in the use of the distributive property.

Guiding Documents
Project 2061 Benchmark
- *Mathematical ideas can be represented concretely, graphically, and symbolically.*

Common Core State Standards for Mathematics*
- *Reason abstractly and quantitatively. (MP.2)*
- *Model with mathematics. (MP.4)*
- *Use properties of operations to generate equivalent expressions. (7.EE.A)*

Math
Distributive property
Multiplication of integers
Expanded notation
Literal notation

Integrated Processes
Using multiple representations
Observing
Recording data
Generalizing
Comparing

Materials
Algebra tiles (see *Management 1*)
Student sheets

Background Information
The article *A New Model for Algebraic Expressions and Equations* on page 5 provides the general background information for this activity.

It is assumed that students are familiar with the rules for multiplying and adding integers and that they have had extensive experience with expressions in different number bases and how such usage relates to symbolic expressions. These activities involve expressions in terms of x.

Three parallel forms are used: manipulatives, representations, and symbolic expressions. The purpose is to show how the same situation can be shown in each of these forms. Because of the one-to-one relationship among forms, students will acquire a broader perspective on algebraic expressions. Understanding is deepened when multiple representations of concepts and processes are provided.

In these activities, the models are built in quadrants I and II. This experience prepares students for the next set of activities in which the models will utilize all four quadrants. The procedure should be studied in reference to the solutions.

Students should be taken through the procedure with several different examples before being asked to work alone or in teams.

Procedure
1. Supply each student or group of students with a set of base-four or base-five algebra tiles. Flats represent x^2 and sticks represent x in this activity.
2. Review why tiles in quadrants I and III are defined as positive and those in quadrants II and IV as negative.
3. The problem identifies the required number and type of manipulatives to be used. Negative components should be distinguished from positive components by marking them in some manner.
4. Instruct students to identify the sets of tiles in each quadrant, including their signs, and record this information in the respective quadrants.
5. Instruct students to write an algebraic expression that includes all of the sets in the four quadrants. The order should be from the largest to the smallest tiles.

6. Review the rules of signs as they apply to addition and subtraction of integers. Use the appropriate sign for each set of tiles. Order should be from the largest to the smallest tiles.

7. Instruct students to form a rectangle with the components making sure that none are moved into a quadrant of a different sign. Be sure students follow the rule of placing the largest pieces in each quadrant closest to the origin and building outward with progressively smaller pieces.

8. Direct them to draw a sketch of the completed model.

9. Instruct students to identify the length, width, and area and write an equation in the form of length x width = area before collecting terms.

10. Have students collect terms in all three representations: in the model by pairing off and removing congruent positive and negative pieces; in the sketch by connecting paired pieces with a two-pointed arrow; and in the algebraic equation by collecting terms.

Connecting Learning

1. What term does not have the same quantity in the algebraic equation after collecting terms as is does in the rectangle, the x-square term, the x-term or the unit term? [x-term]

2. Why doesn't the quantity of the x-term in the algebraic equation after collecting terms match the number of x-strips in the rectangle? [each positive x zeros a negative x]

Solutions

	A	**B**	**C**	**D**	**E**
Positive Components	1 *x*-square 1 *x*-length	1 *x*-square 1 *x*-length	1 *x*-square 2 *x*-lengths	1 *x*-square 3 *x*-lengths	1 *x*-square 3 *x*-lengths
Negative Components	2 *x*-lengths 2 unit squares	3 *x*-lengths 3 unit squares	1 *x*-length 2 unit squares	2 *x*-lengths 6 unit squares	1 *x*-length 3 unit squares
Initial Algebraic Expression	$x^2 + x - 2x - 2$	$x^2 + x - 3x - 3$	$x^2 + 2x - x - 2$	$x^2 + 3x - 2x - 6$	$x^2 + 3x - x - 3$
Drawing of rectangle showing positive/negative pairs					
Algebraic equation before collecting terms: factor x factor = product	$(x-2)(x+1) =$ $x^2 + x - 2x - 2$	$(x-3)(x+1) =$ $x^2 + x - 3x - 3$	$(x-1)(x+2) =$ $x^2 + 2x - x - 2$	$(x-2)(x+3) =$ $x^2 + 3x - 2x - 6$	$(x-1)(x+3) =$ $x^2 + 3x - x - 3$
Algebraic equation after collecting terms: factor x factor = product	$(x-2)(x+1) =$ $x^2 - x - 2$	$(x-3)(x+1) =$ $x^2 - 2x - 3$	$(x-1)(x+2) =$ $x^2 + x - 2$	$(x-2)(x+3) =$ $x^2 + x - 6$	$(x-1)(x+3) =$ $x^2 + 2x - 3$

	F	**G**	**H**	**I**	**J**
Positive Components	1 *x*-square 2 *x*-lengths	1 *x*-square 2 *x*-lengths	2 *x*-squares 2 *x*-lengths	2 *x*-squares 2 *x*-lengths	2 *x*-squares 1 *x*-length
Negative Components	2 *x*-lengths 4 unit squares	3 *x*-lengths 6 unit squares	4 *x*-lengths 4 unit squares	2 *x*-lengths 2 unit squares	4 *x*-lengths 2 unit squares
Initial Algebraic Expression	$x^2 + 2x - 2x - 4$	$x^2 + 2x - 3x - 6$	$2x^2 + 2x - 4x - 4$	$2x^2 + 2x - 2x - 2$	$2x^2 + x - 4x - 2$
Drawing of rectangle showing positive/negative pairs					
Algebraic equation before collecting terms: factor x factor = product	$(x-2)(x+2) =$ $x^2 + 2x - 2x - 4$	$(x-3)(x+2) =$ $x^2 + 2x - 3x - 6$	$(x-2)(2x+2) =$ $2x^2 + 2x - 4x - 4$	$(2x-2)(x+1) =$ $2x^2 + 2x - 2x - 2$	$(x-2)(2x+1) =$ $2x^2 + x - 4x - 2$
Algebraic equation after collecting terms: factor x factor = product	$(x-2)(x+1) =$ $x^2 - 4$	$(x-3)(x+2) =$ $x^2 - x - 6$	$(x-1)(x+2) =$ $2x^2 - 2x - 4$	$(2x-2)(x+1) =$ $2x^2 - 2$	$(x-2)(2x+1) =$ $2x^2 - 3x - 2$

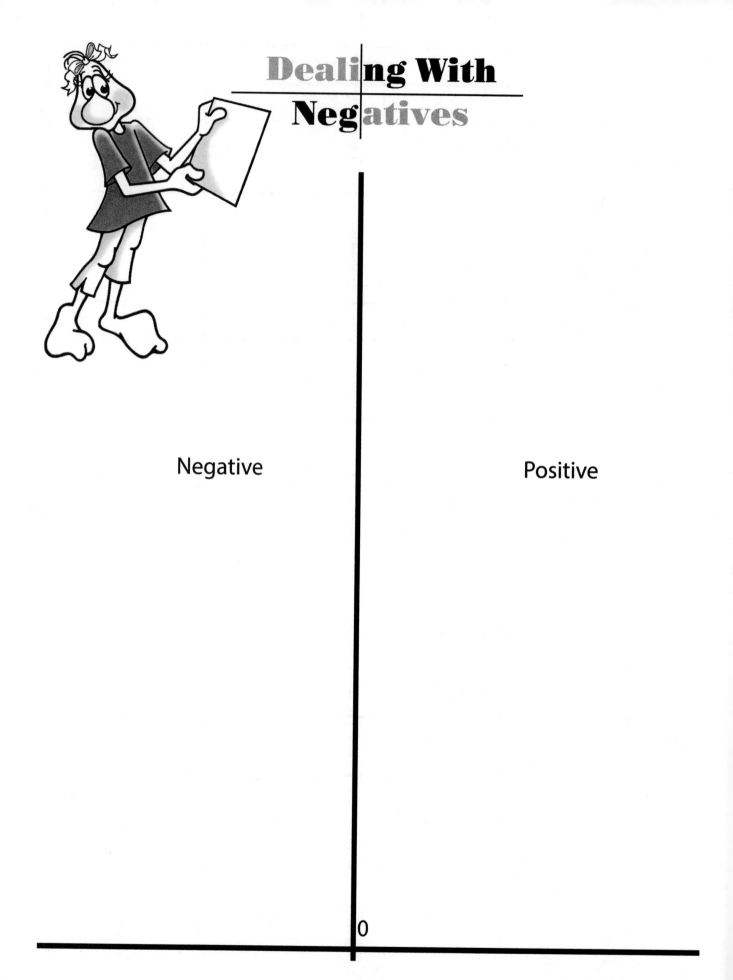

Dealing With Negatives

Negative

Positive

0

Dealing With Negatives

Please do the following:

a. Write an algebraic expression for the sum of the components.
b. Using the components, build a rectangle placing all negative pieces left of the vertical axis and all positive pieces to the right.
c. Write an equation relating the product of the factors and the original algebraic expression.
d. Model collecting terms with the components and re-write the equation with terms collected.

For these activities, you will need these positive components: two 4 x 4-squares and four 1 x 4-lengths, and these negative components: three 1 x 4-lengths and six unit squares.

Since $x = 4$ in these activities, the n-square is 4 x 4 blocks, the x-length is a row of 4 blocks, and the unit square is a single block.

	A	B	C	D	E
Positive Components	1 x-square 1 x-length	1 x-square 1 x-length	1 x-square 2 x-lengths	1 x-square 3 x-lengths	1 x-square 3 x-lengths
Negative Components	2 x-lengths 2 unit squares	3 x-lengths 3 unit squares	1 x-length 2 unit squares	2 x-lengths 6 unit squares	1 x-length 3 unit squares
Initial Algebraic Expression					
Drawing of rectangle showing positive/negative pairs					
Algebraic equation before collecting terms: factor x factor = product					
Algebraic equation after collecting terms: factor x factor = product					

MULTIPLICATION THE ALGEBRA WAY

© 2012 AIMS Education Foundation

Dealing With Negatives

Please do the following:

a. Write an algebraic expression for the sum of the components.

b. Using the components, build a rectangle placing all negative pieces left of the vertical axis and all positive pieces to the right.

c. Write an equation relating the product of the factors and the original algebraic expression.

d. Model collecting terms with the components and re-write the equation with terms collected.

For these activities, you will need these positive components: two 4 x 4 x-squares and four 1 x 4 x-lengths, and these negative components: three 1 x 4 x-lengths and six unit squares.

Since x = 4 in these activities, the x-square is 4 x 4 blocks, the x-length is a row of 4 blocks, and the unit square is a single block.

	F	G	H	I	J
Positive Components	1 x-square 2 x-lengths	1 x-square 2 x-lengths	2 x-squares 2 x-lengths	2 x-squares 2 x-lengths	2 x-squares 1 x-length
Negative Components	2 x-lengths 4 unit squares	3 x-lengths 6 unit squares	4 x-lengths 4 unit squares	2 x-lengths 2 unit squares	4 x-lengths 2 unit squares
Initial Algebraic Expression					
Drawing of rectangle showing positive/negative pairs					
Algebraic equation before collecting terms: factor x factor = product					
Algebraic equation after collecting terms: factor x factor = product					

Dealing With Negatives

Connecting Learning

1. What term does not have the same quantity in the algebraic equation after collecting terms as is does in the rectangle, the x-square term, the x-term or the unit term?

2. Why doesn't the quantity of the x-term in the algebraic equation after collecting terms match the number of x-strips in the rectangle?

Moving Into Four Quadrants

Topic
Distributive property

Key Question
How can the use of the distributive property involving positive and negative integers be modeled on the coordinate plane?

Learning Goals
Students will:
- become acquainted with modeling the distributive property involving integers using all four quadrants;
- build an association between a model, its representation in a sketch, and its parallel algebraic expression; and
- develop a mental image of progessive steps in the use of the distributive property.

Guiding Documents
Project 2061 Benchmark
- *Mathematical ideas can be represented concretely, graphically, and symbolically.*

*Common Core State Standards for Mathematics**
- *Reason abstractly and quantitatively. (MP.2)*
- *Model with mathematics. (MP.4)*
- *Use properties of operations to generate equivalent expressions. (7.EE.A)*

Math
Distributive property
Multiplication of integers
Expanded notation
Literal notation

Integrated Processes
Using multiple representations
Observing
Recording data
Generalizing
Comparing

Materials
Algebra tiles
Student sheets

Background Information
The article *A New Model for Algebraic Expressions and Equations* on page 5 contains the general background information for this activity.

Just as in previous activities, it is assumed that students are familiar with the rules for multiplying and adding integers and have had extensive experience with expressions in different number bases and how that relates to symbolic expressions. These activities involve expressions in terms of x.

Three parallel forms are used: manipulatives, representations, and symbolic expressions. The purpose is to show how the same situation can be shown in each of these forms. Because of the one-to-one relationship among forms students will have acquire a broader perspective on algebraic expressions. Understanding is deepened when multiple representations of concepts and processes are provided.

In these activities, the models are built in all four quadrants. The procedure should be studied as shown in the solutions.

Students should be taken through the procedure with several different examples before working alone or in teams.

Procedure
1. Supply each student or group of students with a set of base-four or base-five algebra tiles. Flats represent x^2 and sticks represent x in this activity.
2. Review why tiles in quadrants I and III are defined as positive and those in quadrants II and IV as negative.
3. The problem identifies the required number and type of manipulatives to be used. Negative components should be distinguished from positive components by marking them in some manner.
4. Instruct students to identify the sets of tiles in each quadrant, including their signs, and record this information in the respective quadrants.
5. Instruct students to write an algebraic expression that includes all of the sets in the four quadrants. The order should be from the largest to the smallest tiles.
6. Review the rules of signs as they apply to addition and subtraction of integers. Use the appropriate sign for each set of tiles. Order should be from the largest to the smallest tiles.

7. Instruct students to form a rectangle with the components making sure that none are moved into a quadrant of a different sign. Be sure students follow the rule of placing the largest pieces in each quadrant closest to the origin and building outward with progressively smaller pieces.

8. Instruct students to draw a sketch of the completed model.

9. Instruct students to identify the length, width, and area and write an equation in the form of length x width = area before collecting terms.

10. Have students collect terms in all three representations: in the model by pairing off and removing congruent positive and negative pieces; in the sketch by connecting paired pieces with a two-pointed arrow; and in the algebraic equation by collecting terms.

Connecting Learning

1. For which rectangles did the number of x-strips not match the coefficient (quantity) of the x-term in the product? [F, H, I]

2. In which quadrants were the x-strips in rectangles F, H, and I? [quadrant I and quadrants II or IV]

3. Why doesn't the coefficient of the x-term in the product always match the number of x strips in the illustration? [each positive x zeros a negative x]

Solutions

	A	B	C	D	E
Positive Components	1 x-square 1 unit square	1 x-square 2 unit squares	1 x-square 4 unit squares	1 x-square 3 unit squares	1 x-square 6 unit squares
Negative Components	2 x-lengths	3 x-lengths	4 x-lengths	4 x-lengths	5 x-lengths
Initial Algebraic Expression	$x^2 - 2x + 1$	$x^2 - 3x + 2$	$x^2 - 4x + 4$	$x^2 - 4x + 3$	$x^2 - 5x + 6$
Drawing of rectangle showing positive/negative pairs					
Algebraic expression for: factor x factor = product	$(x-1)(x-1) =$ $x^2 - 2x + 1$	$(x-1)(x-2) =$ $x^2 - 3x + 2$	$(x-2)(x-2) =$ $x^2 - 4x + 4$	$(x-1)(x-3) =$ $x^2 - 4x + 3$	$(x-3)(x-2) =$ $x^2 - 5x + 6$

	F	G	H	I	J
Positive Components	2 x-squares 1 x-length	2 x-squares 3 unit squares	2 x-squares 2 x-lengths	2 x-squares 3 x-lengths	2 x-squares 3 unit squares
Negative Components	4 x-lengths 2 unit squares	5 x-lengths	2 x-lengths 2 unit squares	4 x-lengths 6 unit squares	7 x-lengths
Initial Algebraic Expression	$2x^2 + x - 4x - 2$	$2x^2 - 5x + 3$	$2x^2 + 2x - 2x - 2$	$2x^2 + 3x - 4x - 6$	$2x^2 - 7x + 3$
Drawing of rectangle showing positive/negative pairs					
Algebraic expression for: factor x factor = product	$(2x+1)(x-2) =$ $2x^2 - 3x - 2$	$(2x-3)(x-1) =$ $2x^2 - 5x + 3$	$(2x-2)(x+1) =$ $2x^2 - 2$	$(2x+3)(x-2) =$ $2x^2 - x - 6$	$(2x-1)(x-3) =$ $2x^2 - 7x + 3$

Moving Into Four Quadrants

I Positive

II Negative

III Positive

IV Negative

Moving Into Four Quadrants

Please do the following:
a. Write an algebraic expression for the sum of the components.
b. Using all the components, build a rectangle placing all negative pieces in the second or fourth quadrant and all the positive pieces in the first and third quadrants.
c. Write an equation relating the product of the factors and the original algebraic expression.
d. Model collecting terms with the components and rewrite the equation with terms collected.

For these activities you will need these positive components: one 4 x 4 x-square and six unit squares, and these negative components: five 1 x 4 x-lengths.

Since $x = 4$ in these activities, the x-square is 4 x 4 blocks, the x-length is a row of 4 blocks, and the unit square is a single block.

	A	B	C	D	E
Positive Components	1 x-square 1 unit square	1 x-square 2 unit squares	1 x-square 4 unit squares	1 x-square 3 unit squares	1 x-square 6 unit squares
Negative Components	2 x-lengths	3 x-lengths	4 x-lengths	4 x-lengths	5 x-lengths
Initial Algebraic Expression					
Drawing of rectangle showing positive/negative pairs					
Algebraic expression for: factor x factor = product					

Moving Into Four Quadrants

Please do the following:

a. Write an algebraic expression for the sum of the components.
b. Using all the components, build a rectangle placing all negative pieces in the second or fourth quadrant and all the positive pieces in the first and third quadrants.
c. Write an equation relating the product of the factors and the original algebraic expression.
d. Model collecting terms with the components and rewrite the equation with terms collected.

For these activities you will need these positive components: two 4 x 4 x-squares, three 1 x 4 x-lengths, and three unit squares, and these negative components: seven 1 x 4 x-lengths and six unit squares.

Since x = 4 in these activities, the x-square is 4 x 4 blocks, the x-length is a row of 4 blocks, and the unit square is a single block.

	F	G	H	I	J
Positive Components	2 x-squares 1 x-length	2 x-squares 3 unit squares	2 x-squares 2 x-lengths	2 x-squares 3 x-lengths	2 x-squares 3 unit squares
Negative Components	4 x-lengths 2 unit squares	5 x-lengths	2 x-lengths 2 unit squares	4 x-lengths 6 unit squares	7 x-lengths
Initial Algebraic Expression					
Drawing of rectangle showing positive/negative pairs					
Algebraic expression for: factor x factor = product					

Moving Into Four Quadrants

Connecting Learning

1. For which rectangles did the number of x-strips not match the coefficient (quantity) of the x-term in the product?

2. In which quadrants were the x-strips in rectangles F, H, and I?

3. Why doesn't the coefficient of the x-term in the product always match the number of x strips in the illustration?

The AIMS Program

AIMS is the acronym for "**A**ctivities **I**ntegrating **M**athematics and **S**cience." Such integration enriches learning and makes it meaningful and holistic. AIMS began as a project of Fresno Pacific University to integrate the study of mathematics and science in grades K-9, but has since expanded to include language arts, social studies, and other disciplines.

AIMS is a continuing program of the non-profit AIMS Education Foundation. It had its inception in a National Science Foundation funded program whose purpose was to explore the effectiveness of integrating mathematics and science. The project directors, in cooperation with 80 elementary classroom teachers, devoted two years to a thorough field-testing of the results and implications of integration.

The approach met with such positive results that the decision was made to launch a program to create instructional materials incorporating this concept. Despite the fact that thoughtful educators have long recommended an integrative approach, very little appropriate material was available in 1981 when the project began. A series of writing projects ensued, and today the AIMS Education Foundation is committed to continuing the creation of new integrated activities on a permanent basis.

The AIMS program is funded through the sale of books, products, and professional-development workshops, and through proceeds from the Foundation's endowment. All net income from programs and products flows into a trust fund administered by the AIMS Education Foundation. Use of these funds is restricted to support of research, development, and publication of new materials. Writers donate all their rights to the Foundation to support its ongoing program. No royalties are paid to the writers.

The rationale for integration lies in the fact that science, mathematics, language arts, social studies, etc., are integrally interwoven in the real world, from which it follows that they should be similarly treated in the classroom where students are being prepared to live in that world. Teachers who use the AIMS program give enthusiastic endorsement to the effectiveness of this approach.

Science encompasses the art of questioning, investigating, hypothesizing, discovering, and communicating. Mathematics is a language that provides clarity, objectivity, and understanding. The language arts provide us with powerful tools of communication. Many of the major contemporary societal issues stem from advancements in science and must be studied in the context of the social sciences. Therefore, it is timely that all of us take seriously a more holistic method of educating our students. This goal motivates all who are associated with the AIMS Program. We invite you to join us in this effort.

Meaningful integration of knowledge is a major recommendation coming from the nation's professional science and mathematics associations. The American Association for the Advancement of Science in *Science for All Americans* strongly recommends the integration of mathematics, science, and technology. The National Council of Teachers of Mathematics places strong emphasis on applications of mathematics found in science investigations. AIMS is fully aligned with these recommendations.

Extensive field testing of AIMS investigations confirms these beneficial results:

1. Mathematics becomes more meaningful, hence more useful, when it is applied to situations that interest students.
2. The extent to which science is studied and understood is increased when mathematics and science are integrated.
3. There is improved quality of learning and retention, supporting the thesis that learning which is meaningful and relevant is more effective.
4. Motivation and involvement are increased dramatically as students investigate real-world situations and participate actively in the process.

We invite you to become part of this classroom teacher movement by using an integrated approach to learning and sharing any suggestions you may have. The AIMS Program welcomes you!

Get the Most From Your Hands-on Teaching

When you host an AIMS workshop for elementary and middle school educators, you will know your teachers are receiving effective, usable training they can apply in their classrooms immediately.

AIMS Workshops are Designed for Teachers
- Hands-on activities
- Correlated to your state standards
- Address key topic areas, including math content, science content, and process skills
- Provide practice of activity-based teaching
- Address classroom management issues and higher-order thinking skills
- Include $50 of materials for each participant
- Offer optional college (graduate-level) credits

AIMS Workshops Fit District/Administrative Needs
- Flexible scheduling and grade-span options
- Customized workshops meet specific schedule, topic, state standards, and grade-span needs
- Sustained staff development can be scheduled throughout the school year
- Eligible for funding under the Title I and Title II sections of No Child Left Behind
- Affordable professional development—consecutive-day workshops offer considerable savings

Call us to explore an AIMS workshop
1.888.733.2467

Online and Correspondence Courses
AIMS offers online and correspondence courses on many of our books through a partnership with Fresno Pacific University.
- Study at your own pace and schedule
- Earn graduate-level college credits

AIMS Program Publications

Actions With Fractions, 4-9
The Amazing Circle, 4-9
Awesome Addition and Super Subtraction, 2-3
Bats Incredible! 2-4
Brick Layers II, 4-9
The Budding Botanist, 3-6
Chemistry Matters, 5-7
Concerning Critters: Adaptations &
    Interdependence, 3-5
Counting on Coins, K-2
Cycles of Knowing and Growing, 1-3
Crazy About Cotton, 3-7
Critters, 2-5
Earth Book, 6-9
Earth Explorations, 2-3
Earth, Moon, and Sun, 3-5
Earth Rocks! 4-5
Electrical Connections, 4-9
Energy Explorations: Sound, Light, and Heat, 3-5
Exploring Environments, K-6
Fabulous Fractions, 3-6
Fall Into Math and Science*, K-1
Field Detectives, 3-6
Floaters and Sinkers, 5-9
From Head to Toe, 5-9
Getting Into Geometry, K-1
Glide Into Winter With Math and Science*, K-1
Gravity Rules! 5-12
Hardhatting in a Geo-World, 3-5
Historical Connections in Mathematics, Vol. I, 5-9
Historical Connections in Mathematics, Vol. II, 5-9
Historical Connections in Mathematics, Vol. III, 5-9
It's About Time, K-2
It Must Be A Bird, Pre-K-2
Jaw Breakers and Heart Thumpers, 3-5
Looking at Geometry, 6-9
Looking at Lines, 6-9
Machine Shop, 5-9
Magnificent Microworld Adventures, 6-9
Marvelous Multiplication and Dazzling Division, 4-5
Math + Science, A Solution, 5-9
Mathematicians are People, Too
Mathematicians are People, Too, Vol. II
Mostly Magnets, 3-6
Movie Math Mania, 6-9
Multiplication the Algebra Way, 6-8
Out of This World, 4-8
Paper Square Geometry:
    The Mathematics of Origami, 5-12
Popping With Power, 3-5
Positive vs. Negative, 6-9
Primarily Bears*, K-6
Primarily Critters, K-2
Primarily Magnets, K-2

Primarily Physics: Investigations in Sound, Light,
    and Heat Energy, K-2
Primarily Plants, K-3
Primarily Weather, K-3
Probing Space, 3-5
Problem Solving: Just for the Fun of It! 4-9
Problem Solving: Just for the Fun of It! Book Two, 4-9
Proportional Reasoning, 6-9
Puzzle Play, 4-8
Ray's Reflections, 4-8
Sensational Springtime, K-2
Sense-able Science, K-1
Shapes, Solids, and More: Concepts in Geometry, 2-3
Simply Machines, 3-5
The Sky's the Limit, 5-9
Soap Films and Bubbles, 4-9
Solve It! K-1: Problem-Solving Strategies, K-1
Solve It! 2nd: Problem-Solving Strategies, 2
Solve It! 3rd: Problem-Solving Strategies, 3
Solve It! 4th: Problem-Solving Strategies, 4
Solve It! 5th: Problem-Solving Strategies, 5
Solving Equations: A Conceptual Approach, 6-9
Spatial Visualization, 4-9
Spills and Ripples, 5-12
Spring Into Math and Science*, K-1
Statistics and Probability, 6-9
Through the Eyes of the Explorers, 5-9
Under Construction, K-2
Water, Precious Water, 4-6
Weather Sense: Temperature, Air Pressure, and
    Wind, 4-5
Weather Sense: Moisture, 4-5
What on Earth? K-1
What's Next, Volume 1, 4-12
What's Next, Volume 2, 4-12
What's Next, Volume 3, 4-12
Winter Wonders, K-2

Essential Math
Area Formulas for Parallelograms, Triangles, and
    Trapezoids, 6-8
Circumference and Area of Circles, 5-7
Effects of Changing Lengths, 6-8
Measurement of Prisms, Pyramids, Cylinders, and
    Cones, 6-8
Measurement of Rectangular Solids, 5-7
Perimeter and Area of Rectangles, 4-6
The Pythagorean Relationship, 6-8
Solving Equations by Working Backwards, 7

* Spanish supplements are available for these books. They are only available as downloads from the AIMS website. The supplements contain only the student pages in Spanish; you will need the English version of the book for the teacher's text.

For further information, contact:
AIMS Education Foundation • 1595 S. Chestnut Ave. • Fresno, California 93702
www.aimsedu.org • 559.255.6396 (fax) • 888.733.2467 (toll free)

Duplication Rights

No part of any AIMS publication—digital or otherwise—may be reproduced or transmitted in any form or by any means—except as noted below.

Model of Learning

Real World
Manipulative
Doing

Number
Abstract
Writing

Counting
Measuring

Graph
Representational
Illustrating

Length of Rope
0   1
Number of Knots

Generalizing
Formula
Abstract
Thinking

$L(k) = 20 - 3k$

MODEL OF LEARNING
Math and Science

The Model of Learning is a foundational component of AIMS lessons. It consists of four environments in which we learn about our world. These environments are represented by four geometric figures: a circle, a triangle, a square, and a hexagon.

An AIMS lesson will start with a *Key Question*. It is this question that leads students into an encounter with the four environments of learning.

For example, the Key Question for an activity might be: "How does the length of a piece of rope change if you tie knots in it?"

The circle corresponds to the real world. It involves *doing* something with or to concrete objects. This environment emphasizes the use of sensory input, involving *observing, touching (taking apart/putting together), smelling, hearing, tasting.* Observations here can be qualitative or quantitative (counting and measuring). The use of multiple senses causes activity in the parts of the brain where that type of information is processed, thus establishing or reinforcing mental connectors.

The activity would begin in the circle with students tying knots one by one in a piece of rope and measuring the length of the rope each time another knot is tied.

The triangle represents the abstractions of *reading* and *writing*. It is the real world symbolized in words and numbers. Meaning is attached to these abstractions because of what was done in the circle. This environment consists of recording numbers that result from the counting and measuring. It may involve writing a description of what was observed. Reading the rubber band books found in the AIMS Science Core Curriculum or the cartoons found in the Math Modules are included in this environment. Students work here when they read their textbooks or do the computation exercises.

In the knot-tying example, the students would next move to the triangle where the number of knots and the length of the rope are recorded in a table.

The square represents *picturing* or *illustrating* the real world. Graphs, diagrams, drawings, or an isometric drawing are examples of the pictures that can be used. The square might simply involve picturing what was recorded in the triangle. It could also be an illustration of an object or event that occurred while students were working in the circle environment. Both

of these situations are constructed from student input; however, the drawing or graph or other illustration could be one that is imposed upon a student for interpretation purposes.

Relating to the example of tying knots, the student might move from the triangle to the square to construct a graph of the relationship between the number of knots in the rope and the length of the rope. Or the student could have moved directly from the circle to the square to construct the graph.

Finally, the hexagon represents *thinking, analyzing, generalizing, creating formulas, hypothesizing,* and *applying.* What did we find out? What does it mean? Is there a relationship? What is it? Is there a formula? To what other situations might this apply?

Students in the example might look back at the table and be asked, "By how much does the length of the rope change each time another knot is tied? How does this show up in the graph?" Students might note that the graph is a line. "What is the slope of the line? How does the slope show up in the table? What would be the length if there were 10 knots in the rope? How about if there were n knots in the rope? How could you find the length without measuring? Could you write this as an equation?"

The arrows pointing to and away from each of the environments suggest the importance of moving back and forth between the environments. To generalize the relationship between the number of knots and the length of the rope required going back and forth between the hexagon, the square, and the triangle. Once the formula is found, it can be applied to another situation that might be posed in the circle.

Perhaps the most important thing the Model of Learning does is to remind us of the four learning environments and help us think about how to structure an activity or lesson so that students are constructing and reinforcing the concepts and relationships. This is accomplished by moving back and forth between these environments. The triangle is where students too often spend most of their time in school activities. While this environment is not unimportant, the other environments give students something other than the tip of their pencil with which to think. AIMS activities are designed around these environments and make moving back and forth between these environments natural and meaningful.